Newton's Gravity

SECOND EDITION

HOW A HIDDEN HEURISTIC USED BY NEWTON HAS
COMPROMISED MODERN PHYSICS

LUTHER L. NAYHM, PHD

Published by Tarkas Press

Copyright © 2019 Luther L. Nayhm

All rights reserved. No part of this work may be reproduced in any form without the written permission of the author.

ISBN: 9781729339916

All great truths begin as blasphemies

> — George Bernard Shaw

A new scientific truth does not triumph by convincing its opponents and making them see the light, but rather because its opponents eventually die, and a new generation grows up that is familiar with it.

> — Max Planck

The first principle is that you must not fool yourself — and you are the easiest person to fool.

> — Richard Feynman

I think I can safely say that nobody understands quantum mechanics.

> — Richard Feynman

A great deal of my work is just playing with equations and seeing what they give.

> — Paul A. M. Dirac

Contents

Prefaces i
Introduction 1
Chapter 1—The Center of Mass and Modeling 7
Chapter 2—The First Anomaly: Orbital Angular Momentum 15
Chapter 3—a Hidden Omission 39
Chapter 4—Further Discussions on Newton's Gravity 62
Chapter 5—General Conclusion on Newton's Gravity 81
                    Appendices 91
Appendix 1—Orbital Angular Momentum....................................... 93
Appendix 2—Newton's General Gravitational Model 105
                    References 113
References 115

Preface to the Second Edition

The second edition improves some of the descriptions and cleans up typos as well as enhances the mathematical models and discussions in Chapter 4. Chapter 4 is focused on the mutual forces on objects embedded within extended distributions of mass. The original choice of the normalizing function in these models was not the best for showing how the generalized Newtonian gravitational model's outcome differs from the outcomes when using traditional Newtonian point-mass gravitational model. In particular, for extended disks of matter, the force on an object imbedded within that discoidal distribution takes on the appearance of the types of forces that require dark matter within the galaxy.

In addition, most calculations were made using improved modeling techniques provided by members of the Mathematica User Group, who introduced these improved techniques and reproduced some of the results presented in Chapters 3 and 4. These changes also improved the clarity of the images of the results of the numerical analysis used in this book.

One additional issue that was addressed was to better understand the differences in results from using various coordinate systems to describe the shapes used to find the mutual gravitational interactions between objects and then to separate these variations into their root causes. Variations arise from the particular algorithms used in making the numerical calculations either in cylindrical or spherical coordinate systems, and these variations must be distinguished from real variations in the results that are a result of using the general Newtonian gravitational model versus using the point-mass gravitational model.

Preface to the First Edition

This book is the first in a series of four books, and in this first book we discuss Newton's gravity and how we have overlooked a glaring error that changes many representations of how objects interact gravitationally. We discuss how and why this error may have occurred and the consequences of this error.

This preface is verbose and includes general information relevant to all the books in the series, including why the series exists. In addition, this preface and the prefaces to all the books contain material important both in understanding the context of each book within the series and in describing the approach that is consistent throughout the set. Hence, the verbosity.

The first two books in the series are introductory physics books, whereas the third book is about a specific classical propulsion technology, which is both a novel technology and an extension to, as well as an example of, an existential technology. The discussion pertaining to the existential nature of the technology stimulated additional research into why the new principle for a propulsion system that should have been identified contemporaneously with the development of rockets was not recognized until the present.

Overall, the series represents a history-of-science journey into a domain of science considered complete and unimpeachable. Under normal circumstances, the history-of-science scholar would read what others have written and respond with some analysis on what those prior descriptions signify, if anything. But a "reporter's" approach does not work for this series of books, because, as it turns out, the history of many of the physics discussed in the series are a result of mistakes, omissions, and self-deceptions. Most people who research and write about the early work in physics are true believers in the efficacy of those physics, at least as they form a pathway toward the modern versions embraced by the consensus. We acknowledge that errors and omissions may have occurred in the very earliest scientific attempts, but we rely on these being caught and corrected by the time the modern paradigm for these physics or science become the basis for moving forward or, as often happens, for things to stagnate. What this series of books shows, however, is that the modern consensus was reached too quickly, freezing out further progress in these areas of science.

Not being content to simply review and record what others said about certain fundamental physics, I reviewed the foundational work produced before the consensus was reached and identified many instances in which work was left undone. The paradigms were established leaving holes that, by the nature of paradigms, rejects the possibility that such errors or omissions might exist. Since the research for this book clearly shows that the holes, omissions, and incompletions exist, this series has become not just a microscope putting the paradigms under high magnification, the series becomes a scalpel for opening these paradigms up for further reviews and criticisms.

The first book on Newtonian physics was stimulated by the resistance to the new classical propulsion technology, which is not proscribed but is contrary to certain well-established beliefs, which turn out to be heuristics. Heuristics are simplifications, generalizations, or rules of thumb that seem to be good enough for all practical purposes, but the issue emerged as to whether they are as good as we think, though the real issue is that many of these heuristics are not recognized as being heuristics. An observation that was made early in my career of an unrecognized anomaly in an analytical modeling effort using Newtonian physics identified a starting point for investigating whether Newtonian physics as we understand it was complete.

My research into whether we were interpreting Newtonian physics correctly or completely led to the discovery of a vast number of heuristics within physics, many of which have altered our perceptions and beliefs in what is scientifically and technically allowed. In developing his models for orbits and for classical gravity, for instance, Newton used the center of mass, a concept he inherited from antiquity, but as he used it, the center of mass inadvertently became a heuristic. By using the concept of center of mass in the ways that he did in developing his orbital and gravitational models, he inadvertently bequeathed to us a slavish belief in these heuristics rather than in the exact physics.

As another example, in developing his gravitational model, Newton made an additional error, which occurred because he inadvertently used the point-mass heuristic to simplify his model without knowing or proving that his model as he developed it and as we use it is a heuristic. In plain terms, Newton did not do what he said he was going to do in developing his gravitational model and did not validate what he did do as being the same as he intended to do. Newton applied inductive reasoning in accepting the universality of the center of mass as used in antiquity. We show that such an inductive extrapolation is wrong, which raises the issue as to whether inductive reasoning is ever legitimate within the context of science or technology other than to supply a guess as to which steps to take in some area of research.

We discuss the origin and consequences of the gravitational model errors in considerable detail, because if the accepted gravitational model is a heuristic, what is the correct model, and what are the consequences of having used the heuristic model for the past three hundred years? The discussions begin using simple algebra and logic, including using Newton's own words from his *Principia Mathematica* to show that his gravitational model is a heuristic. Since general relativity reduces to the Newtonian gravitational model in the absence of mass, then general relativity is also a heuristic.

More importantly, a new generalized Newtonian gravity model was developed and applied to a variety of scenarios. These scenarios consisted of using spherical and cylindrical objects in various combinations and orientations to identify where and how the general model might impact our understanding of other physics and sciences. In addition, when the model was used to investigate the internal forces on a test mass within a volume of mass, the results were significantly different from those found with the point-mass model.

My approach to the two physics books was to make them readable to a wide variety of people who would be interested in the topics but who might lack either the historical context from which the science emerged or the algebra I use in my analyses, which are key elements in my approach. In the first two books, I start many chapters with a brief historic discussion of the topic of the chapter. The purpose is to help guide the discussion through both the physics and how, why and when the omissions or incompletions arose. For the most part, I also relegate the higher-level mathematics and modeling to appendices.

I also supply some fundamental references or at least enough that the reader can navigate through the more popular or common sources of information and cull what is superficial or even incorrect. Simply stated, if I quoted hundreds of sources, not only would most of them be irrelevant, no one would follow up on these and, more importantly, most people I want to read this book do not have access to these references, though the non-scientist and scientist alike are often dazzled by copious reference for their own sake. If I cannot find a reference on the Internet, I usually excluded it from my list. In more scholarly books, or at least those pointed toward a more academic audience, a tidal wave of references is often a type of pandering and a false indication of goodness, thoroughness, and competence…and even stamina, which I must admit to admiring.

References also protect authors from claims of plagiarism, even though ideas are often lifted from other's efforts. It is the verbatim copying of text that causes problems, yet ideas are often simply taken and massaged using new language. Any quote can be reparsed if the subject matter is understood well enough, so plagiarism is usually a lazy person's offense. I seldom quote but I do attribute.

My approach was to let my work stand on its own merits. Beyond a certain point that was quickly reached, the work is unique and must stand on its own. References mainly supply a learning path for those who are not subject matter experts in the topics under discussion. Consequently, for this series of books, my references are specifically directed toward various learning paths, given the breadth of the audiences with which I wish to communicate. On the other hand, often the omission of a particular reference will fuel criticism by a subject-matter expert without their having understand the author's work. In the subject matter for this book, ultimately there are no real subject-matter experts, only poseurs.

I did need to supply sufficiently accessible references so that the reader could dig into what I was saying to the extent that they could…if they were so motivated…in order to render a judgement on the soundness of my logic and discoveries. In general, an Internet search over a named individual, topic, or event will yield enough information to understand the named person, topic or events' contributions or significance to the narrative. I make no attempts at supplying directly referenced documents or citations in these instances.

None the less, we are coached to distrust Wikipedia, though we should be coached to treat everything with the same skepticism, even information on university web sites and certainly on government web sites and in all the books written about modern physics. The value of Wikipedia is to establish a vocabulary from which more in-depth searches can be conducted and from which key historical or current R&D

practitioners' research can be identified. Wikipedia also contains some surprisingly good references in many articles, which can also help guide a further and more in-depth search. Whenever a link disappears, if the topic and names of people are available, the link can usually be bypassed and new ones identified. This is also true of many university or government links which simply disappear as sites are consolidated into new sites or abandoned entirely.

I would suggest that the reader also become familiar with Google Scholar, since many good references are supplied in a downloadable form. However, much there is also referenced to refereed journals for which one must either subscribe or have access to a university library...I have both but chose not to include most of these as references though I did access many such documents in researching this book, hundreds, in fact.

I also perused the arXiv.org databases managed by Cornell University, but many of those documents need to be approached with caution, since they are not stringently refereed as posted, where the affiliation and prior productivity of the author is deemed sufficient for establishing reliability and quality. Unfortunately, many documents are unreadable, and their content should be suspected. Yet occasional dissertations and "teaching" papers show up that can be very useful to the determined researcher. It is interesting that certain documents that have little commercial value but historically have much to offer are also posted in arXiv and can be extremely useful in supplying lesser known references or insights. In some cases, a published article of some interest is also available in arXiv as a preprint and I do reference these types of documents, since they have a higher level of vetting relative to most arXiv documents that are not subsequently published, plus they can be downloaded and read by anyone.

I reference many accessible documents, but there are also other sources of material that are accessible at reasonable cost. Some I reviewed and others I supplied from reputation and my own experiences with these sources. One source is the Dover Publishing library of classical texts on a wide variety of topics. Another is the Cambridge University Press imprints of historical books, some of which are inexpensive. Another source is the Schaum's Outlines, which while not quite the "Cliff Notes" of physics and mathematics, cover a wide range of topics from introductory to advanced that are useful in rekindling ones' skills and knowledge in math and physics. Additionally, my original "go to" series as an undergraduate was the Feynman Lectures, though these are not introductory as originally intended and are somewhat dated now, fifty years after their first introduction.

And, finally, specific topics in analytical mechanics were derived from two well-known but expensive text books authored by Goldstein., Poole., and Safko and by Kleppner and Kelenkow, though I have many such books in my library. Much of the physics I used from these two sources can be found in the Dover books and in the Schaum's Outlines but you must look carefully in these sources to find that material.

As a final note, I need to address the issue of why should anyone bother to read this book or any of the others in the set? Who am I to take on these topics and claim to have uncovered hidden truths? Foremost, I am an observer and a skeptic, plus I have sufficient education, experience, and knowledge to understand the details of the topics about which I am writing. The real issue is that I pursued the research into Newton and Einstein and their physics in a way apparently never attempted. Had anyone with my experience and skepticism undertaken similar research, they would have made the same discoveries as I have.

Progress in science typically occurs one baby step at a time. It takes a considerable number of small steps to arrive at some large step that is often an epiphany if not a consolidation of prior baby steps. Throughout this book, I make a reference to the butterfly effect, in which the beating of a butterfly's wings at some point can cause a significant deviation in the course of progress in any field. It was first used to describe the sensitivity of weather models to minor changes in the initial conditions to the models, though the general concept is at least a century older. What is also true is that often the wing beats lead to disastrous outcomes. The point is that just because the butterfly's wings change the trajectory of some future events does not mean that these were all leading to positive outcomes. Bad ideas lead to bad predictions and bad outcomes.

We started the series of books with Newton, because Newtonian physics is the best known of the physics discussed in the series, and the familiarity of this subject matter creates an opportunity to more easily understand the format and approach used in the complete series of books. We also start with Newton, since as the King said to Alice, "Begin at the beginning...."

Introduction

If you jumped here without reading the preface, please go back and read it before continuing. And yes, I am nagging you. The preface is an introductory essay that sets the tone for the discussions in this book. While the current volume is presented as a standalone volume, it is really part of a sequential set. The general theme of the set is that we have misused and misinterpreted certain physics, starting with Newton's hidden heuristic. These missteps have propelled us into and beyond Einstein's relativity along a path that has become in many instances both dead ends and fabrications. Moreover, we have formed a rigid belief that our fundamental physics is inexorably true and complete, which has further inculcated a sociology in which we deny that we have the rigid beliefs that we hold while simultaneously dissuading any deep dives into these fundamental physics.

The discussion starts with Newton, but the issues begin even earlier than Newton and are associated with what he believed and why. Newton believed in the point-mass and center-of-mass concepts in physics. The question of "why did he believe in these ideas" is simple: he had no reason not to believe in them. There were no compelling observations that indicated they were wrong or incomplete except, perhaps, one observation by Huygens, which is discussed later. However, this same honest appraisal cannot be maintained moving beyond Newton, because there are reasons not to believe in certain fundamental concepts that form the foundation of modern physics. But this is getting ahead of the story.

One thing Newton did believe in is our modern classical view of how extended objects interact via gravity, with each mass element in each object simultaneously acting on and being acted upon by each mass element of the second object, which is a viewpoint rejected by the hardcore general relativist but remains the classical description of Newtonian gravitational attraction between extended objects. However, Newton also believed that reducing one of the objects to a point mass at

Introduction

the center of mass of an object was equivalent to characterizing the complete object as it interacted with a second object. His mathematics purports to prove this.

But Newton's point-mass and center-of-mass mathematics are not the same as the way he described two extend objects interacting, which is also how we describe these interactions now. Unnoticed and unremarked upon is that Mach's ideas, which are discussed later, on the origins of inertia are identical to Newton's ideas on how objects interact gravitationally. Newton, however, left out one step in showing how he proved his gravitational law, which was to prove that the simultaneous interaction as he and we believed it occurs was equivalent to his reducing one of the objects to a point mass. For three-hundred years we have not noticed this lack of rigor. As it turns out, the two approaches for modeling gravity are not equivalent, though they can be close, especially for spheres but not so much for other shapes and mass distributions within objects.

An issue that is addressed repeatedly in this book is that it is only in the modern era of computers and powerful analytical computers models that certain of the models that will be discussed could be accurately evaluated. Yet, countering this is that the mutual interactions between certain regular shapes for objects do create simple analytical models. The required calculus has been available for several hundred years, yet these special interaction geometries were never examined. This adds credence to the observation that Newton's development within the *Principia* was both a heuristic and a non-sequitur...Newton did not do what he said he was going to do and never acknowledged that he was using a heuristic because he did not recognize the point-mass concept as limited in any ways.

The gravitational model error or "oversight" has had consequences that have rippled through physics for the past three hundred years. My contention is that this single oversight and how it occurred and how it has persisted has promoted other oversights and out-right mistakes and proscriptions that have impacted our abilities to generate new technologies. One such technology, discussed in detail in the volume *A Novel Propulsion System*, is erroneously proscribed. This technology has potential existential consequences, which means that any country that perfects this technology can and likely will in some ways dominate everyone else. All these issues revolve around the mistake in Newtonian gravity, which has infected physics in many deleterious ways.

Therefore, there are two mistakes or blunders, as I characterize them in the final book in the series, in Newton's gravity law. One was Newton's belief in the center of mass and the other is was his ignoring his own statement of what he was trying to do in setting up his gravity model: he did not do what he said he was going to do and then proceeded to tell us that he had done what he said he would do. This is common in propaganda, though by all evidence, this was an honest mistake on Newton's part, unlike the usual applications of propaganda.

We also observed that he made this same type of error in developing the model for orbits. Later we discuss that it was the observation that the point-mass approach to orbital angular momentum was a heuristic that led to the observation that Newton used the same assumptions in his gravitational model. When either the gravity model or the orbital equations are rigorously developed by explicitly avoiding the use of the center of mass, these models are not the same as the point-mass models: they are close but not precisely the same. But, and this is the main issue, no one wants to talk about what is a straight forward breakdown in logic, rigor, and completeness and, I might add, competence. What will emerge as the series of books progresses is that academic physics more than any other scientific discipline is held in excessive and pathological thrall to its paradigms.

Breakdowns in logic occur over and over in science and have their underpinnings in not doing what we are saying and not saying what we are doing. In Newton's case, it was likely that he simply did not recognize that he had made this error, whereas others have used this technique as a way of obfuscating what are otherwise vague, problematic, or contentious ideas. A propaganda style is intended to mislead the reader, either as conscious or unconscious efforts at obfuscation. The reluctance of academics to talk about what Newton had done…or not done…propelled me forward on a simultaneous quest to understand why the reluctance as well as to trace the consequences within physics of Newton's errors. This quest brought me face to face with the pernicious qualities of paradigms and the sociology that sustains these paradigms in the face of evidence that they are false paradigms.

Gordin and Popper have commented on the fact that theories are often held on to long after they should be discarded. They do not go far enough in explaining why that is so. Kuhn's definition of a paradigms, which is a variable concept that is far from fixed, is often used to justify the status quo, whereas these same paradigms are used to justify a society that, while claiming to be progressive, is actually regressive in the sense that we can never go back and review the origins to the paradigm. This is clearly the academic paradigm associated with Newtonian physics, which is claimed to be unimpeachable and, as such, beyond being reviewable. That Newtonian physics is, in fact impeachable, is the hypothesis examined in this book.

The following discussions are only a summary of the voluminous quantity of research that has occurred in understanding Newton and his scientific and non-scientific works. My intention here is to point out those areas in which Newton was working that are germane to the discussion of where the omissions in physics occurred in which Newton did not complete his analysis in a critical research area. Understanding how the omission or incompletion occurred required understanding, in part, what he did know and what he believed.

Newton's best-known and most influential writing was his *Philosophiæ Naturalis Principia Mathematica*, or simply *Principia*, though scholars have delved extensively into his other writings, including letters to peers and various commentaries by his contemporaries on both Newton and their common interests. Otherwise, we can also glean much by understanding the status historically of the natural philosophy during Newton's time. As will be pointed out, some ideas had existed perhaps for millennia before Newton was active and he had no reason not to believe that these ideas were true…so he incorporated some of them into his research work.

The scientists in the 1600s, which includes Newton and his peers such as Hooke, Leibniz, and Huygens, among others, were incredibly clever and brilliant when you consider what they accomplished with the tools at their disposal. They were observers of the highest order. Many of these people despised one another and intellectually interacted reluctantly if at all. A short paper by Nobel Laureate Sheldon Glashow, "The Errors & Animadversions of Honest Isaac Newton", supplies an entertaining summary of the fact that geniuses can be real pieces of work. More than that, Glashow's article points out that science, like history, is often written by the winners of various disputes and intellectual competitions. What this book shows is that often the consensus by scholars is not a strong indicator of correctness.

Glashow's article further points out that we often have a superficial or idealized view of the contributions of certain of the pioneers in physics or in any field. Sometimes history makes erroneous attributions or even erases contributions from the record. In many cases the contributions by others are coopted by a more dominant personality. Despite Glashow's exposé of Newton, for purposes of this book, it is irrelevant whether Newton came by his ideas independently or by purloining them from others. What has come down to us from that time forms our core beliefs in our basic physics. The discussions in this book are concerned with the physics and not with the people and their behavior or contributions except to supply a general context for the discussions.

Any errors attributable to Newtonian physics are errors supported by 300 years of true believers. Consequently, what Newton knew and when or why was a common set of beliefs that we might just as well attribute to Newton as to Hooke or anyone else. And, if we are to believe Weinberg's and Kuhn's rationale, we can attribute the errors in understanding Newton's physics and mathematics to all the active scientists of the era who had a hand in modernizing Newton's physics but who did not fully understand or recognize what they were reading in the *Principia*.

Chapter 1—The Center of Mass and Modeling

The center of mass is a concept introduced early in physics curricula in what is called analytical mechanics. It is fundamental to the study of statics, which has its roots in antiquity, but Newton introduced and discussed this simple idea in his modeling of dynamics, especially orbital dynamics. Orbital dynamics are also discussed in introductory university studies in analytical mechanics, whose roots have been essentially declared complete and correct for a century. Significantly, our modern approaches to modeling orbital dynamics contain ideas unfamiliar to Newton, such as energy and momentum conservation. None the less, we still use the center of mass just as Newton described and used it. As used in statics, the center of mass is not a heuristic.

In "The Revolution That Didn't Happen," Nobel Laureate Steven Weinberg describes how our understanding of the works of the pioneers, such as Newton, is often based on the evolution and refinement of the work of those pioneers rather than on their exact work. What we call Newtonian physics is actually post-Newton physics. Our modern approaches to these pioneering physics is associated with our better understanding of these physics through the accumulations of small butterfly wing beats driving our knowledge and techniques forward.

Sometimes, though, as with the center of mass, incomplete ideas are also driven forward, and physicists have proven reluctant to question too deeply those ideas that will be shown to be heuristics, because heuristics are wonderfully easy to use and rely upon. I am always skeptical of ideas that are too simple, too linear, and seemingly too universal. Weinberg speaks to this obliquely by describing how difficult it is for physicists to understand the former paradigms when they are working in a more modern paradigm. The issue may be more related to the culture within academic physics that enforces paradigms at the expense of constantly questioning these same paradigms.

Little new research has occurred in classical analytical mechanics in nearly a century, though the applications have continued into the modern era in understanding orbital mechanics for specific scenarios, such as for spacecraft and planetary objects, and research over the past half century or so is continuing into non-linear mechanics and stability theory. Progress, if we wish to characterize how mechanics is taught now versus a hundred years ago, is primarily associated with "modernizing" the mathematical tools and approaches used for what is essentially the same physics. We have, ironically, become fixated on "rigor" while not ensuring that we have historically been rigorous.

The orbital mechanics Newton developed are hardly recognizable in how we approach orbital dynamics today. Newton deduced the nature of the orbital kinematics using geometry and algebraic descriptions of orbits by employing the early recognition of the existence of centrifugal forces arising from constrained motion along an arc, a notion supplied in part and indirectly by Huygens, Hooke and others. Glashow's article shows that the real evolution in the understanding of dynamics by Newton was likelier much slipperier and more convoluted than the above description might indicate.

Newton showed that the empirical observations and deductions of Kepler could only be met by the specific interaction forces called central forces. Central forces are the mutual interactions of two objects in which, as Newton described it, the centers of mass of the interacting objects cause the mutual forces, attractive in the case of gravity, to be along the line connecting the centers of mass. Again, Glashow identified Hooke's fingerprints all over this description. Using models for these central forces, Newton showed how Kepler's three laws of planetary motion could be met by a force that varied as the inverse-square of the range between two point masses, which, in Kepler's scenario, was between a planet and the sun. However, as Glashow points out, maybe the above attribution should go to Hooke…who knows? Newton also showed how specific variation in these mutual forces would perturb the orbits in specific ways, such as producing a precession of the orbital ellipse.

It was only much later that actual orbits and equations for conic sections describing these orbits were produced in a more modern and recognizable algebraic form, which required the development of more modern calculus and notations, which did not occur until the mid-1700s. Though the ways in which these new forms of calculus could be used to model dynamics did not occur until the early to mid-1800s. In the modern era, we have well developed methods of building dynamic models that use basic calculus and the calculus of variations that employ energy and angular momentum conservation, all concepts and mathematics that emerged in the mid-to-late 1700s up through the mid-1800s. The progress was slow and steady but not as neat and seamless as our textbooks might seem to indicate.

Newton's approach to his modeling was based on a concept much older than him, which is the idea of the center of mass. A brief survey of pre-Newtonian research shows that Archimedes was likely the first or one of the first to articulate the concept. None the less, the notion of viewing the center of mass as containing all the mass of an object in a dimensionless point called a corpuscle was already established in Huygens's time as Huygens's experimented with pendula. Kepler also used that idea in his analysis of orbits.

However, careful experimental efforts in what we now call dynamics were in short supply in the 1600s. Consequently, how and when the idea of the center of mass, corpuscles, and point masses became a recognized tool of the trade for dynamic modeling is not clear but certainly predated Huygens and, as noted above, was likely articulated by Kepler in his analysis of planetary orbits, for which the physical size or shape of the orbiting objects was not of any consideration. Since detailed dynamic observations were only beginning to be made in and about the sixteenth century, it might be surmised that the idea for the center of mass containing the total mass of an object is certainly over four hundred years old, perhaps even millennia old, and we still use that belief in modern analysis.

Consequently, Newtonian mechanics and dynamics are built on a foundation much older than Newton himself. Archimedes experimented in what are called statics, which is the study of how forces on objects cause the object to spin or experience torques, among other things. Archimedes efforts were not precisely analytical as we understand the term now, but he recognized the scaling involved, such as what happens as a lever length is increased. The same observational models employing geometry and trigonometry were widely and successfully used in the ancient world, as the monumental engineering feats of the ancients' can attest. We can find the physical center of mass of an object, for instance, by simply hanging it from various points on its edges and noting where the extensions of the hanging lines cross within the object. This of course finds the center of gravity which is essentially the center of mass. The center point for mass could be considered as containing all the mass of the object in working with static modeling, such as for levers and torques.

Newton had worked out the models for motion under a central force such as gravity contemporaneously with the development of his gravitational theory, which, per Newton, was being discussed among his peers. The gravitational law for point masses would have to have existed at the time Newton developed his analysis of Kepler's laws of motion. However, Newton was apparently reluctant to write his modeling efforts down for others to see and critique. Under some pressure to "publish", Newton finally produced his magnum opus, the *Principia Mathematica*, or more formally his *Philosophiae Naturalis Principia Mathematica*. All his dynamic modeling and analysis as well as his approach to these models and his gravity model were recorded in the *Principia Mathematica*, and the interested reader can find the *Principia* plus copious historical material on-line within The Stanford Encyclopedia of Philosophy.

The Stanford Encyclopedia of Philosophy referenced above became my gold standard for understanding and researching the works and thinking of many famous researches and their approaches to their work, especially in the areas of the philosophy and history of science. The issue was to discover what they knew or believed and when they knew it or believed it and how those beliefs were reflected in their interpretations of their work.

The *Principia* was a compendium and description of both existing knowledge that was being discussed by Newton and his peers and the new ideas and extensions that were uniquely Newton's. Still, per Glashow, we must be wary of inaccuracies and attempts to divert attention within what may have been a compendium of Newton's work intended to project rigor where rigor was lacking. I am skeptical, despite these demeaning practices, that there are any purposeful fabrications per se in the *Principia*: Newton wrote about what he believed in. That in some ways he may also have been a poseur is a separate issue.

The modern model for motion in the presence of a central force can be found in any modern analytical mechanics text worth its salt, and the introductory texts are the clearest. More advanced texts are focused on mathematical rigor as much as the physics. However, Newton's approach to the dynamic models for the orbit of the planets used a graphical approach that did not result in algebraic descriptions of the motion. The reason for looking at these models is that the topic of interest for the first beating of the butterfly's wings is orbital angular momentum, and the orbital angular momentum is a critical element to any orbital modeling, at least for modern approaches to modeling orbital dynamics.

Newton and his peers only had a general idea of how to represent such concepts as angular momentum. In fact, Newton and his peers would not have been familiar with some of the elementary concepts presented in introductory mechanics books today, and angular momentum was one such concept. Newton's approach was essentially to treat all mass in an object as a dimensionless corpuscle located at the center of mass of the object, and his efforts at finding the mutual force between extended objects would have relied on the idea of corpuscles, an approach that had taken root prior to Newton's work.

The Stanford Encyclopedia of Philosophy entry for Newton supplies a brief discussion relating to angular momentum in the discussion of Newton's laws of motion. In that discussion in the last paragraph, it is noted that Huygens had observed that the motion of a pendulum with two separate masses did not oscillate as if the total mass was located at the center of mass of the two masses on the pendulum arm. The arc of a pendulum can be considered as part of a trajectory, and Huygens had observed that for angular motion, the center of mass motion experiences a perturbation away from what would be expected when the second mass was added. Huygens's anomaly would have been known to Newton but apparently did not stimulate any particular discussions at the time, and the relationship of the small arc of a swinging pendulum to a segment of an orbit was not noted, likely because no specific equation of motion had been developed by Newton to describe an orbit that specifically contained a quantity known as angular momentum. In fact, Huygens's model for the motion of the pendulum was itself a heuristic he developed through trial and error. In the limit of small swinging arcs, his model is very accurate and is still used today, though it is now derived from first principles using modern ideas and physics.

The above historical descriptions introduce the idea of mathematical modeling, which owes its existence to the approaches that Newton introduced in arriving at his descriptions for orbits, among other physics he pursued. The idea is that we can describe events mathematically, such as describing how objects accelerate and the paths along which objects move. More generally, legitimate quantifiable science and engineering require the development of mathematical models, often referred to by the philosophers of mathematics and science as representations. These models describe how the scientific and engineering activities will progress and what results can be expected. Unless the reader has a firmer understanding of what is meant by scientific or engineering modeling, the context for the physics omissions will not be as obvious as needed to understand the impact of these omissions.

What many scientific representations supply, especially in physics, is a validated description of something that is physical. Something that will emerge later and which is important in the third book is the description of centrifugal and Coriolis forces, which occur on an object when it is in a moving frame, such as rider on a carousel. Without belaboring the point here, our models based on validation shows that in certain spinning systems, the two forces on a mass, the centrifugal and the Coriolis forces, differ by a factor of two and act in directions perpendicular to one another. These are precise observations based on experimental verifications. Consequently, we can rely on these results and models when we are modeling the behavior of mass in such systems, such as in modeling a centrifugal pump moving fluids. These are not arcane observations passed down from master to apprentice in secretive notebooks, these are "facts" anyone can rely upon. And that, to me, is the true value of our analytical models, though sometimes these models contain bad or speculative physics with no observational basis.

It is important that the approach to modeling be clear, since in the next volume on relativity, the discussion relating not only to special relativity but also to theoretical modeling in general will demonstrate that a lack of understanding of the significance of measurement models will be identified as a root cause of false claims and assertions on exactly what is or has been theoretically modeled. Or, to say it another way, if a theoretical model is not founded in the requirements of how to make a measurement necessary to validate the model, the models will likely be wrong or wrongly interpreted. Within special relativity both conditions exist.

Using a straight forward example from my past, I will describe one of the roles modeling plays in modern scientific and engineering activities. In the example I will use, we wanted to make a measurement of the transmission properties of the atmosphere under specific circumstances. To frame the required measurements or experiments, we needed to develop scenarios for how the measurements would proceed, and the mathematical models supplied a description of these scenarios. From these descriptions, we implemented the experiments and recorded the outcomes we expected and from the outcomes deduced the properties of the materials introduced into the measurement path, though ultimately the goal was to understand the reverse, which was how the materials impacted the measurements.

Without belaboring the details of our approach and rationale for making the measurements, without a description of both the physics and the process, we are not performing an experiment per se in the modern sense, since to simply make the measurement without understanding what is going on means we cannot properly interpret our findings…if any. Sometimes no result was expected or the absence of a result is itself useful information. And, sometimes, our interpretations are wrong because we did not fully understand the totality of the physics and processes that were used in doing an experiment.

The above description was supplied merely to show that the idea of modeling is essential in understanding how to make measurements and how to interpret the results which are measured. This is the underpinning to the scientific method in which we expect to record, quantify, and duplicate our results to understand how and why the measurement process proceeds as it does. Otherwise, we would still be in the pre-scientific era of natural philosophy where, as with alchemy, we simply try things out and look for outcomes and from that try to recognize repeatable patterns. Newton was in the forefront of driving the transition from natural philosophy into science and his early development of models was fundamental to that transition.

But, to be fair, pre-Newtonian scientists did make observations and they did record how they did their experiments and then recorded the outcomes of those experiments. What was missing was the analytical descriptions that quantified the experiment and its results. For instance, when Galileo apocryphally dropped objects from the Leaning Tower, he knew that they would fall to the ground. He had deduced that the acceleration of gravity would obey certain laws such as the distance being proportional to the time squared, though he had no way of expressing this algebraically. He would also observe that changing the weight did not change the drop time. All in all, his observations would be recorded qualitatively with no real basis for predicting anything other than the fall time of an object was independent of the object's weight. He may have observed that size of objects with the same weight might affect the fall time, and this might lead him to suspect that air resistance was involved. But the point is that without analytical models, his interpretations of his observations were limited. As with Newton, some of these dynamics were being discussed by others and in some instances long before Galileo undertook his own analysis. Newton gets the credit for some physics because he was the first to analytically describe quantitatively what other had already described correctly but qualitatively.

As for the center of mass, in the next chapter we describe why the center of mass is a heuristic in dynamic modeling, including why Huygens's pendulum's oscillation rate did not behave as he expected when the additional mass was added. When an approach used in dynamic modeling is a heuristic we would expect that approach to fails to support observations under certain circumstances. What we show is that for the most part the observations do support the center of mass in dynamic modeling, but that is because the perturbations are simply too small either to be observed or to have an impact on practical technology. We will quantify the impact of avoiding the used of the center of mass to show that, while the center of mass is a heuristic, it simply does not matter except within certain academic physics, where the consequences of the center of mass being a heuristic change how we interpret certain scientific measurements and observations.

Chapter 2—The First Anomaly: Orbital Angular Momentum

Orbital angular momentum is called the first anomaly in this book. It is only first in the sense that it was the first anomaly I observed that stimulated me to question whether we were modeling the angular momentum correctly. The real first anomaly would be our use of the center of mass without exploring its true dynamic characteristics. Later I describe a very simple model that would have exposed both the center of mass and the angular momentum, specifically orbital angular momenta, as being heuristics and, therefore, being the source of the same anomaly. That the orbital angular momentum was a heuristic is true because it used the center of mass, which is a heuristic in dynamic modeling.

Early in my career I was working under contract to such organizations as DARPA and NASA, among other government R&D organizations. I was looking at orbital dynamics and how we might better understand the signatures of orbiting and re-entry objects or vehicles. By signatures is meant that we were looking for unique observational characteristics from tracking the movement of these objects, where we would be observing the objects using optical or radar sensors in both active and passive modes of operation. Active refers to sending out a beam of radiation that would reflect off an object and subsequently be detected. Passive refers to simply looking at the object as it was being irradiated by natural sources of radiation from the sun or, in the infrared, emitting thermal radiation. The goal was to develop unique measurable observables that were sufficiently unique to improve the detection, tracking, and identification of these objects. The inverse was that knowledge of such signatures would allow us to develop ways of masking those same signatures in our own orbital or reentry vehicles.

In this chapter, we "get into the weeds" on a topic that will be shown to have limited practical impact. There is academic impact but little impact on technology. The reason for insisting on the details in this chapter is that it sets a tone for what is described in the rest of the book. We show that our unquestioning reliance on the simple point-mass ideas in dynamic modeling has led to inventing new physics to explain deviations or perturbations that the simpler heuristic models did not explain.

The First Anomaly: Orbital Angular Momentum

Solving the orbital equation of motion produces a solution that is a description of a conic section, which consists of various open and closed curves such as circles, ellipses, parabolas, and hyperbolas, which are the trajectories of an object moving under the influence of an inverse-square force such as gravity and which arises from the mutual interaction between two objects. The actual physics is for both interacting objects to be orbiting around the barycenter of the system, where the barycenter is the center of mass of the system as well as a focus point of the curves describing an orbit or trajectory, which are the conic section, and we won't talk about that any more. But of most interest in this chapter is the orbital angular momentum identified as is p_θ. (The θ in p_θ is a subscript and not a multiplier.) We show below where p_θ occurs in the actual equation for orbital motion.

The discussion of orbital angular momentum is Newtonian physics without being something on which Newton directly performed modeling and analyses. The orbits are a result of the application of Newton's laws of motion. However, models of orbits introduce the notions of both the center of mass and how it is a heuristic whose use leads to other heuristics. Because the use of the center of mass in many other models is so prevalent, we need a strong understanding of why the center of mass is sometimes a heuristic and how it forces any other models in which it is used to be a heuristic.

Many people over the past century and a half have studied orbital motions and the shapes and stabilities of those orbits as a function of the law governing the attraction (or repulsion) between objects. The inverse-square law for attraction supplies the unique solutions that are defined as conic sections. The inverse-square law in Newtonian gravity is given as $F = G M m/r^2$, where F is the mutual force between the two orbiting objects of point masses m and M, G is the universal gravitational constant, and r is the separation between the centers of mass of the two objects. Newton had also investigated the consequences of a non-inverse-square force law and deduced that Kepler's laws could only be met for planetary motion if the force were exactly an inverse-square force. Finally, the most important parameter in the orbital equations is p_θ, the angular momentum, which is the source of the first anomaly, which also contains what might be called the hidden primordial assumption of the center of mass being a point containing the total mass of an object at the object's physical center of mass.

It is at this point that we employ a simplification that may or may not prove to be another hidden heuristic. In finding the orbital equation, an observation is made that two objects orbiting one another can be described by separating the dynamics into two pieces. One piece describes the common center of mass or barycenter's motion and the other piece describes the direct interactions between the co-orbiting objects. Consequently, if one knows how the center of mass or barycenter of the two orbiting masses is moving, this motion is additive to the collective motion of the two objects about their common barycenter. We can see what this means by considering the Earth orbiting the sun. The whole solar system is in motion around the galaxy, and the Earth is orbiting the sun. We can, to first order, separate these two motions. Later we will discuss that maybe we cannot or should not do that, but for now, to first order we can separate these two motions.

At this point we need to alert the reader about some other consequences and unresolved issues surrounding what we have just discussed. On the one hand, we have an unresolved issue in how and why we never questioned the center of mass and its use in Newtonian dynamics. On the other hand, there is a more pernicious issue in that no one wants to believe that from the small beginnings in this chapter, we can show that almost all our current Newtonian mechanics and much of modern physics are simply heuristics or completely wrong, with the exact dynamic models in hiding and waiting to be discovered.

Over time, all these heuristics can be exceeded, which mean that the models incorporating the heuristic do not predict what we measure. When this happened with Newtonian dynamics, we invented new physics rather than going back and investigating the roots to the existing physics. This has created problems in physics and technology that will be discussed throughout the remainder of this book.

The starting point in addressing the above issues is orbital mechanics. Some modern approaches to orbital dynamics are more elegant than others, but all result in the same model. One approach uses a conceptually simple idea that is related to the conservation of energy, an idea that was introduced qualitatively long ago but was only articulated mathematically in ~ 1830s using mathematics developed in the late-1700s by Lagrange in his reformulation of mechanics. The approach uses an advanced equation called the Euler-Lagrange equation, which was first developed in the mid-1700s as an element of the then-new calculus of variations. The Euler-Lagrange equation in mechanics uses a function called L, the Lagrangian, which is simply T - U, where T is the kinetic energy and U is the potential energy. L is easy to find but solving for the equations of motion for orbits using *L* in the Euler-Lagrange equation can be more challenging. The Lagrangian was not "invented" by Lagrange but was identified by Hamilton later and then named after Lagrange, since the Lagrangian allowed Hamilton to perfect his own further reformulation of Lagrangian mechanics into the modern forms we use today.

The key is that the angular momentum of the orbit falls naturally out of the Euler-Lagrange equation as an important contributor to the shape of an orbit called the orbital eccentricity. However, the concept for the potential energy, *U*, did not evolve until the early 1800s with work performed by Faraday involving electric charge. Consequently, the approach to orbital dynamics using a Lagrangian likely evolved simultaneously with the approaches developed by Hamilton in the 1830s for modeling dynamical systems. Note that in the above description and those in Wikipedia, we can see that the modern forms for modeling dynamics evolved over the course of a century by incorporating incremental refinements in the understanding of physics and the adoption of mathematics developed for different purposes. These are our butterfly wing beats in action.

What is often also overlooked is that once the orbital equation is found, the actual orbiting mass is a reduced mass, which is an algebraic relationship between the two orbiting masses. The reduced mass is given as $\mu = m_1 m_2/(m_1 + m_2)$ for the two masses m_1 and m_2. Using this simplifying transformation, the two-body problem becomes a one-body problem in finding the orbital dynamics. What is really happening is that the two objects are co-orbiting their common center of mass, which we describe next, but instead of describing both objects orbiting, we can reduce our focus to only one object that has a modified mass...at least for the point-mass models.

From a practical perspective, the first-order orbits hold when M>>m, so that the smaller object is nearly orbiting the larger object, which holds well enough for planetary orbits during Newton's time and artificial satellite orbits in modern times. And, more precisely, the orbits must be found for the co-orbits of both objects about the barycenter of the system, which is the center of mass for the system of two objects. An analogy would be a barbell. We can balance the barbell at one point on the cross bar...the center of mass. If we add weight to one end, the balance point shifts toward the larger mass. This balance point in an orbital system is called the barycenter of the system and is the common orbital point. If we were to stably spin an unbalanced barbell around some pivot or balance point, this point would be the barycenter of the barbell.

We do not need to go into the details of finding an orbit or even into the various simplifying assumptions, since the goal here is not to actually calculate an orbit but simply to state how and why the angular momentum comes into play. To move the discussion forward, we need to dig into the details of exactly what the angular momentum is and how it enters the above equation of motion. As noted above, Newton did not need or have a strong working knowledge of angular momentum, and Huygens's anomalous observations relating to the period of a pendulum did not stimulate any scrutiny of the idea of the center of mass not being an appropriate way to describe motion.

Huygens's model for a pendulum used a center of mass approach, in which the mass of the bob does not enter the model and the location of the center of mass of the bob from the pendulum's pivot point determines the length of the pendulum's actual swing arm. Only the length of the pendulum arm and the acceleration of gravity determine the period of a pendulum for small swing arcs. When Huygens added another "bob" or second mass to the pendulum arm at a midpoint, he did not understand why the period had changed to something that he was not expecting, only that it did not match a "virtual" pendulum length taken from the pivot to the center of mass location for the two bobs.

It is likely that the use of the center of mass to describe the mass of an object in motion was a customary practice at the time, which was before Newton's active period. Thus, any description of angular momentum would likely have used a center of mass location to describe the effective point at which all the mass of an object was considered to be located. We have for the most part maintained this view point to the present. However, the following discussion will show why Huygens's pendulum's period changed.

Formally, the orbital angular momentum is given in modern notation as **L = r X p**, where bold letters represent vectors and **X** is the cross-product operator. The cross product between two vectors is just the sine of the angle between the two vectors. The parameter **L** is not to be confused with the Lagrangian defined previously. To avoid any confusion, the orbital angular momentum within the orbital equation of motion will be given as p_θ. Since linear momentum equals m v, and **v**, the velocity, is a vector, we have that linear momentum is a vector and, consequently **L = p_θ** is a vector, too. The magnitude of the angular momentum is then given as p_θ = r m v sinθ, where r

is the distance from the orbital barycenter or center of curvature to any mass element *m* moving with a speed v in the direction θ relative to the vector **r**.

The operator **X** is used to find $\sin\theta$ in the expression for p_θ. What the cross product and resultant sine do is calculate the component of the velocity vector that is perpendicular to the vector **r**, or alternatively and equivalently, to calculate the component of **r** that is perpendicular to **v**, either-or. In mechanics, this is called the lever arm against which a force is pushing to cause a torque which causes spin. (In the approximation used in this chapter, the distance **r** is the distance from the barycenter and the centers of reduced mass of the single orbiting object, which does not impact the general conclusions. The barycenter is just a point in space.)

Under the assumption that all the mass in an object can be considered located at a point within that object called its center of mass, which is the only concept Newton understood, we have then that $\sin\theta = 1$ and $p_\theta = m_{tot} r v = m_{tot} r^2 \omega$, where $v = r\omega$ and where ω is called the angular rate of the object moving along a curve. The angular rate is referenced to the center of curvature of the arc being followed. This form makes it easy to find the angular momentum for an orbiting object, which then allows the actual orbit of that object to be found.

As Newton understood the behavior of an object in orbit, the distance from the center of curvature to the center of the object, which is **r**, was a vector that was perpendicular to the tangent to the arc at any and all locations of the object moving along the arc or orbit. Newton would not have been familiar with the above description though he may have intuitively understood what was occurring. Despite our knowledge of the limitations in Newton's understanding of angular momentum, we persist in using the center of mass description for the angular motion of an object in motion. However, this simple approach is where the anomalous results were identified many decades ago, which is getting ahead of the discussion.

Such a persistent oversight in some detail of an historical work is not unusual. **Astrophysicist Subrahmanyan Chandrasekhar**, a Nobel Laureate in physics, pointed out in 1995 that the consequences to one of the theorems in Newton's *Principia* related to calculating orbital dynamics had been overlook for three centuries and has since been studied in relationship to orbital stability. It is not surprising that other issues may have been overlooked. It is surprising that Chandrasekhar, in his delving into the *Principia*, failed to catch the inconsistency in the point-mass and gravitational models, likely because he was looking for something subtle, and the issue with the point-mass gravitational model is not so subtle but is so deeply imbedded in our belief in the point-mass gravitational model that it is essentially off the table for deeper review. If someone other than Chandrasekhar had pointed out the error he did find, they would likely have been ignored. None the less, the fact that Chandrasekhar obviously stared the discrepancy in the gravity model in the eye and did not see it is symptomatic of a persistent mind-set among physicists that will be pursued later in trying to understand how such obvious errors were overlooked by so many people for so many years.

Going back to the orbital equation, the significance of p_θ can be better seen by rewriting the orbital equation in terms of the shape and dimensions of the orbital path, which for a closed orbit is an ellipse. Open orbits such as parabolic and hyperbolic orbits also exist but are characterized by an object not actually being in an orbit, and the paths of the object relative to the central force in the system, such as the sun and some comets or asteroids, are simply one-time trajectories and do not repeat. Rewriting the orbital path, we have $1/r = (1+\epsilon \cos\theta)/(a(1-\epsilon^2))$, where a is the semi-major axis of the ellipse (one half the long dimension of the ellipse) and ϵ is the eccentricity of the ellipse, which defines an ellipse's characteristic oval. The above equation for r holds for all trajectories, open or closed and all elliptical or circular orbits, depending on the magnitude of ϵ which in turn depends on the magnitude of the total energy…kinetic plus potential…of the moving object.

Re-writing the equation for the eccentricity, we have that $\epsilon = (p_\theta^2/r_o G m^2 M) - 1$. We arrive at this by noting that we need to have some starting point in the orbit for measuring θ, so the location we pick for setting $\theta = 0$ is taken as the point of closest approach of the orbiting object to the barycenter, which is the periapsis of the elliptical orbit, and we set $r = r_o$ to represent this distance. Therefore, setting $\theta = 0$ in the above equation, $\cos\theta = 1$ at $r = r_o$.

Elliptical orbits have both closest and farthest orbital points relative to the barycenter of the orbit, which is typically located close to the largest object in an orbiting pair of masses. For orbits about a stellar object, the closest point of approach is called the perihelion after Helios for the sun, though Helios is a specific reference to our sun. For an orbit of a moon about a planet, this same term is periapsis. Periapsis is a modern term coined at the beginning of the space age, but the two terms have the same meaning in terms of an orbit. The farthest orbital point is called aphelion for a stellar orbit and apoapsis for a moon's orbit. Periapsis and apoapsis are considered

The First Anomaly: Orbital Angular Momentum

the generic terms and perihelion and aphelion are specific to orbits around the sun or a star, though, as stated, Helios is specific to the sun. And, of course, we have apogee and perigee for satellite orbits around the Earth, where Gee is a reference to Gaea, the Greek goddess of the Earth. In an overly pedantic way, we can coin other such orbital extrema for Jovian or Saturnian or Martian satellites and so forth.

For all orbits, through the conservation laws, we have that p_θ is taken as a constant, and angular momentum is conserved in an orbit. The conservation occurs because the only force in the system is the central force acting in the direction r and not in the angular direction θ, where such an angular force would change the speed of an object and the angular momentum would not be constant or conserved. Based on the prior discussions, the total angular momentum of the system would be the sum of the angular momenta of both objects as they orbit the barycenter of the system. The above approach is standard within advanced text books, yet there are several assumptions that may indicate that the orbital equation is itself a heuristic and needs much more effort to be cast as a rigorous equation of motion.

Now, here's the question: what do we mean when we say p_θ is conserved or constant? There would not appear to be any way other than to interpret that statement as the angular momentum is fixed for any given object orbiting another object. But, if we look at an exaggerated shape, such as a dumbbell in orbit, we can test the above statement very simply. We employ the description of a standard dumbbell with two equal masses m attached to one another by an essentially massless bar of fixed length d. If we put the dumbbell into an orbit with a barycenter Δ from the center of mass of the orbital system to the center of mass of the dumbbell, which is the center of the bar, we can orient the dumbbell with the bar parallel to and then perpendicular to the vector Δ.

When we use the formal definition of angular momentum, which is **L = r X p,** we can recalculate the angular momentum for each orientation of the dumbbell. Using the point-mass approach, we find that when **d** is parallel to Δ, we have that $L_{cm} = 2\,m\,\Delta^2\,\omega$, which is the classical value of the orbital angular momentum for the center of mass of an orbiting object of total mass *2 m*. In this case $\mathbf{P}_{\theta\,cm}$ is still the angular momentum regardless of the orientation of the dumbbell.

However, when an object is in orbit, its angular motion, ω, is constant for all mass elements within the object, but the velocity of each mass element depends on the distance of that mass element from the orbital barycenter.

Consequently, for a real object, all the mass is not at the center of mass of that object. If we use the exact expression **L = r X p**, we find something different for the orbital angular momentum of a dumbbell. The magnitude of **L** becomes $p_{\theta\,tot} = p_{\theta\,cm}(1 + d^2/4\,\Delta^2)$, assuming that the mass at the ends of the connecting rod are concentrated at the center of mass of the shapes containing the mass, which for standard weights are spheres or disks, and the length d connects those mass points. We can see that this is another assumption that means that the above perturbation is only first order and that there is yet another correction factor with which to contend, which is the same assumption of center of mass for each mass, which we will ignore here.

We can find $p_{\theta\,tot}$ from the above analyses by finding the angular momenta of each mass separately, and then summing these together for the whole orbiting dumbbell. Performing the modeling as described, we find that $p_{\theta\,tot} = m\,\omega\,[(\Delta - d/2)^2 + (\Delta + d/2)^2] = 2\,m\,\omega\,\Delta^2 + m\,\omega\,d^2/2$, which can be simplified to $p_{\theta\,cm} + m\,\omega\,d^2/2$. When we factor out $p_{\theta\,cm} = 2\,m\,\Delta^2\,\omega$ from this expression, we find that $p_{\theta\,tot} = p_{\theta\,cm}(1 + d^2/4\,\Delta^2)$. This form is only first order since we also assumed that the mass of each end of the dumbbell is concentrated at the center of its own distributed mass. To find the exact expression, we need to use the technique discussed in Appendix 1. This was the first modeling efforts I used in an early contract that identified the anomalous term in the orbital angular momentum of an object. Later this simple approach was expanded to include other analytical shapes, which is also discussed in this chapter.

The factored form above for *p*₀ *tot* obscures an important element of the physics. While the factored form shows that the anomaly is related to the ratio of the dimension of the object to its distance from the barycenter of the orbit, the un-factored form shows that the perturbation is a constant perturbation of the point-mass orbital angular momentum. While the point-mass part of the angular momentum shows how the angular rate varies as the distance *Δ* varies, which occurs in any non-circular orbit or path, the perturbation is a constant independent of *Δ* for all path shapes. Thus, the angular momentum in the orbit equation is still conserved. The conservation is less obvious in the factored form.

However, the un-factored form also allows us to identify what the perturbation is. The perturbation is the moment of inertia of the dumbbell for rotation about its center of mass. The moment of inertia is usually labeled *I* and the angular momentum is $p_\theta = I\,\omega$, where *ω* is some spin rate. Therefore, when

The First Anomaly: Orbital Angular Momentum

Huygens added a new mass to his pendulum, the swinging angular momentum was not simply for a single center of mass but, as with the dumbbell, he had two different angular momenta that when added together did not equal the angular momentum found by assuming that the two masses created a single new point mass in a new location. Therefore, the period of the pendulum changed in a way that Huygens did not understand but it foreshadowed the anomalous angular momentum I calculated three centuries later.

The actual rotation for an orbiting dumbbell with a fixed attitude is occurring at the angular rate ω of the orbit, when the axis of the dumbbell is kept parallel to Δ during an orbit, which is added to the inherent spin rate of the object about its center of mass. Thus, the perturbation is associated with a single rotation of the object at its orbital rate…to first order. This will become clearer in the next examples for the perturbation for a sphere and cylinders in orbit.

The perturbation arises, as stated earlier, because the speed or magnitude of the velocity of each point in an orbiting object depends on its distance from the barycenter, and only the angular rate, ω, is a constant for the whole object. Each point on the object has a new velocity. Hence, the center of mass erroneously assumes that the whole object has the velocity of the center of mass. This was the source of the perturbation Huygens measured for the period of a pendulum, which swings in a small arc. Huygens problem is often included as a student problem relating to angular momentum in analytical physics text books, such as that authored by Thornton and Marion, though the relationship of this problem is not associated with Huygens.

When we re-do the math for an orbiting dumbbell with the cross bar along the direction of motion, the two masses are the same distance from the barycenter and have, consequently, the same speed. A vector Δ' from the barycenter to each of the masses m is not perpendicular to the velocity vector, but the small angle just compensates for the increase in the distance from the barycenter to each mass. Therefore, the resulting angular momentum is the point-mass momentum and there is no perturbation.

In a moment, we will look at more typical shapes of some orbiting objects, such as spheres and cylinders. We will assume that these objects have a uniform density, though it is clear that any variations in the density, in the shape, and in the orientation of the object with respect to the barycenter results in a variation in the angular momentum. The consequences of this are that the eccentricity is not constant, because the angular momentum is varying as an object moves along its orbit, whether the object is spinning (or tumbling) or has a fixed direction relative to the stars. When an object is pointing in a fixed direction, then, relative to the object about which the object is orbiting, the object's orientation is changing and, consequently, so is the angular momentum, unless the orbiting object is continuously re-orienting to keep a fixed attitude relative to the barycenter of the orbit. In this latter case, the perturbation is constant and does not vary, and the eccentricity is constant but of a slightly different value than the point-mass angular momentum, because of the additional amount of angular momentum supplied by the single rotation.

We will not go into the detailed analysis of the variation in eccentricity, but we will describe the consequences. First, the shape of the orbit will vary from that predicted using the point-mass approach. If an object is spinning or tumbling, the result would be a slight scallop shape added to the orbit which would repeat with a frequency given by the object's rotational rate. This repeating scalloping would be superposed on top of the actual orbit that is perturbed by the single hidden rotation that occurs once per orbit.

In other words, the single-orbit perturbation to the angular momentum defines the new eccentricity of the actual orbit, and any spinning of the object causing orientation changes supplies a second additive angular momentum perturbation that results in the scalloping motion of the object. That is, it is not that the spinning per se is coupled to the orbital angular momentum but rather it is that the orientation change resulting from the spin that changes the orbital angular momentum of the object.

The object does not have to be orbiting for its trajectory to experience the scalloping perturbation. Any non-spherical or non-uniform object moving under the influence of gravity and spinning about some axis would also have a scalloped motion superposed on its trajectory. The reason for these scallops is that any rotation of the object is causing each mass element to have a varying distance from the orbital or trajectory center of motion. Consequently, while the rotational angular momentum may be conserved, the orbital angular momentum is fluctuating, which impacts the instantaneous orbital eccentricity.

Detailed analysis also shows that the point of closest approach, r_o, would also change, and analysis of the equation for r and the angular momentum show that the object would have small oscillations in both the radial and angular rates during each cycle of the scallop during an orbit. The object would seem to speed up and slow down slightly and to slide or oscillate forward and backward and in and out radially by small amounts during each scallop cycle about some mean position. The mean position would be determined from the mean value of the actual fluctuations. However, a uniform sphere has a fix perturbation that would not vary as the sphere spins, but the orbit and the eccentricity would have a fixed perturbation. In the case of a space telescope that might point in a single stellar direction during an orbit, the scallop would be a single orbit in duration and would show up as a fixed perturbation in the orbital eccentricity.

The good news is that the mean or point-mass orbital parameters would essentially hold even if there are small oscillations of the object's motion and position during each scallop. The bad news is that unless we are aware of the existence of the scallop cycles, any short-term observation of the object would be put into a center of mass orbital equation to predict the actual path or trajectory of the object. The good news is, if the scallop perturbation is too small to measure, which is fortunately usually the case, we would have erroneous radial speed and distance measurements that, while small, could lead to what are commonly known as navigation or tracking errors. For most orbiting objects, but not all, the scallop and angular momentum anomaly are, while present, unobservable. We will discuss the exceptions later.

Given the above discussion, it may be that these variations in orbital eccentricities are only of interest for larger planetary objects. It is possible, though unlikely, that we may be seeing such errors in the tracking of certain Earth-orbit crossing objects, especially very large and oddly shaped objects with significant rotations (or tumbling). These objects are extraordinarily difficult to initially detect. Furthermore, the speeds and angular motion are also difficult to measure. The perturbations from the angular momentum in the presence of the multitude of gravitational perturbations make the detailed calculation of the trajectories of these objects difficult to predict with any long-term precision.

We read in the news that scientists are constantly updating their predictions on exactly how near to Earth an object will come when it crosses Earth's orbit. In addition, it is likely that stability theory incorporating these new perturbations would predict the existence of random variances from orbit to orbit because of the non-constant value for the eccentricity of an orbit. Thus, the prediction of a recurring orbit may experience what appear to be random perturbations such that longer-termed multi-orbit predictions will show unexpected variations. Consequently, a predicted miss of the Earth may subsequently not be a miss.

To be fair, the above constraints to accurate course prediction are extremely difficult with which to contend. For most man-made objects and typical meteorites and most asteroids, the magnitude of the variations in the angular and radial positions and the perturbation on the mean orbital eccentricity are in the centimeter or less range and the angular rate deviations during a scallop are less than milliradians per second. For near-Earth man-made objects, or planetary object of similar dimensions, variations are all in the millimeter and milliradian ranges. While such measurement precision could be made in a laboratory or under very specific and controlled circumstances, the general ability to simply point a laser or microwave radar at moving objects and to collect enough accurate information to determine the characteristics of the scallop perturbations remains non-existent, especially for changes in the angular rates.

From a practical perspective, the center of mass remains a useful method for trajectory predictions, except that the input parameters that are consistent with the scallop perturbations are absent, but using current sensors, the perturbations are simply buried in the signal noise anyway. We know what our sampling and integration times must be to extract such signal variations, but as the signals are weak for more distant objects, we need to use pulse methods that further add constraints to the allowable sampling and integration times. The shapes and motions of the objects introduce their own signal perturbations that are larger than those produced by the scallop perturbations. It is likely that we may never be able to make the types of measurements that are needed in a timely fashion to allow such trajectory predictions to become accurate enough for whatever passes for practical circumstances at some future date.

But, there you are, we have some new tools in the bag of analytical tricks for making such trajectory predictions. Just as importantly, we now know that our text book descriptions are incomplete and that each scenario must be evaluated on its own merits to make sure that the dynamics are correctly handled when using the well-known and easier-to-use heuristics.

We have missed the scallop perturbations because, from a practical perspective, we did not know to look for them. But, and this is the big but, it is in the job description of academic physicists to be thorough and to investigate the impractical and then for the impractical to be made practical by the technologically inclined among us. There is really no excuse not to have looked deeper into such a simple concept as the center of mass. But that is, again, getting ahead of the story.

Suffice to say, we have identified the first butterfly effect, which is that the center of mass is a heuristic and is not precise, and use of the center of mass is an approximation for certain types of models. Now, what does that mean? At least in the angular momentum, we can find the exact value without resorting to a heuristic, but it remains to be shown what that really means and what the more significant consequences to using this heuristic has produced in the ensuing centuries.

To move the discussion forward, believe it or not, we need to dig still deeper into the angular momentum. However, before supplying any more discussion relating to the angular momentum and other consequences of the inaccuracy of using the point-mass approach to find these angular momenta, it is useful to look at the more formal way of making the types of calculations than were used in the dumbbell example. The heavy lifting is relegated to Appendix 1, but the broader results and implications will be discussed in this chapter.

The dumbbell model only required simple algebra and geometry for finding the orbital angular momentum, but the dumbbell shape is a much simpler shape than usually encountered and it is clear from the discussion below and in Appendix 1 that our treatment of the dumbbell was simply a first-order effort as persuasive as it is. Consequently, we need to make a start on how we would look at the angular momentum of uniform objects such as spheres and cylinders, which were chosen because they produce, by and large, analytical models for their angular momenta. Even so, there are certain circumstances in which the models require numerical solutions, and I used Mathematica® in working through numerous examples.

The first useful element of a formal calculation of orbital angular momentum is that we are discussing motion about a point, a center of angular motion or a barycenter. We can draw a vector from the barycenter to the center of mass of an object and use this as an axis of symmetry for our models. I have chosen to call this vector Δ and it is equivalent to the z-axis in cylindrical and spherical coordinates. The total angular momentum is the sum of all the incremental angular momenta from each mass element in some orbiting object. This is more general than the above description might appear, since any object moving in an arc in which some force is acting to cause the arcing motion allows an instantaneous center of curvature to be identified from which the instantaneous angular momentum is found. Thus, a vector $\boldsymbol{\Delta}$ exists for all locations along the arc.

Next, we form a vector from the center of mass of the object to any mass element within the object, and the vector sum of this local vector with $\boldsymbol{\Delta}$ defines the actual path used to find the angular momentum of that mass element. However, we also need to find the vector cross product of this vector with the velocity vector of that mass element. Since the angular motion is common to every mass element within the object, the motion vector is the tangent to the curve the total mass is following, which is perpendicular to $\boldsymbol{\Delta}$. What this says is that the velocity vector can be set to point along the x-axis and the right-hand rule defines the direction of the y-axis. We then define all the vectors in terms of their (x,y,z) coordinates. The incremental angular momentum of the ith mass element is $p_{\theta\,i} = m_i\, r_i\, v_i\, \sin\theta_i$, and to find $p_{\theta\,tot}$ we sum all i^{th} volume increments in the object, which is simply the integral over the volume of the object. The details are shown in Appendix 1.

We can perform the integrals for spheres and for cylinders with various orientations relative to the velocity vectors. When we do this, we find that the orbital angular momentum consists of two parts. One part is the center of mass angular momentum and the second part is a perturbation that is dependent on the dimensions…and mass distribution…of the object. For the uniform sphere, $p_{\theta\,tot} = p_{\theta\,cm}\,(1 + 0.4\, r^2/\Delta^2)$, where r is the radius of the sphere and we have assumed a uniform density within the sphere. The un-factored and constant perturbation term is $p_{\theta\,pert} = 0.4\, \omega\, m\, r^2$, where ω is the orbital angular rate of the sphere. We can allow the mass density to vary, but if the distribution is not analytical, we would find a numerical result and not the exact model as shown above. Thus, if the cylinder is rotating or tumbling with a rate ω_t, we might see scalloped motion with a rate ω_t. If a sphere has asymmetrical mass distributions within it, then we would also see a fluctuation in the orbital angular momentum plus a fixed mean perturbation in the eccentricity that defines the nominal orbit of the object.

For a uniform cylinder with its axis parallel to the vector $\boldsymbol{\Delta}$, we find that $p_{\theta\,tot} = p_{\theta\,cm}\,[1 + (\Lambda^2 + 3\, r_c/12\, \Delta^2]$, where r_c is the radius of the cylinder and Λ is the length of the cylinder. For the cylinder axis pointing in the direction of motion, we have that $p_{\theta\,tot} = p_{\theta\,cm}\,[1 + r_c^2/2\, \Delta^2]$, where now there is no length dependence. For the cylinder orbiting with the velocity vector being a

The First Anomaly: Orbital Angular Momentum

perpendicular bisector of the cylinder, we find that $p_{\theta\,tot}$ depends on the orientation of the cylinder with respect to the vector Δ. If the cylinder were rotating at some rate around the perpendicular bisector, the total angular momentum would fluctuate, again leading to scallops in the orbital path.

If we look at a uniform sphere, the total orbital angular momentum is $p_{\theta\,tot} = m\,\Delta^2\,\omega + 0.4\,m\,r^2\,\omega$. The second factor is simply the angular momentum of a uniform sphere rotating at a rate ω about an axis through the sphere's center of mass. If we calculate the rotation of a sphere about an axis tangent to the surface of the sphere, we have $\Delta = r$ and $p_{\theta\,tot} = 1.4\,m\,\omega\,r^2$. We show a more complete description for the calculation of the orbital angular momentum in Appendix 1.

But what the above model proves is something called the parallel-axis theorem of mechanics. The parallel axis refers to a second spin axis that is parallel to but off-set from the spin axis that passes through the center of mass of an object. Using the concept of the moment of inertia, I, described previously, we have that the moment of inertia of a sphere $I_{sphere} = 0.4\,m\,r_s^2$ and that $p_{\theta\,sphere} = 0.4\,m\,\omega\,r_s^2$, which holds for a uniform sphere, and the calculation of the moment of inertia proceeds exactly as shown for the orbital angular momentum or the angular moment of inertia. If we now put another axis parallel to the center of mass axis but displaced a distance d from that axis, we find that the new moment of inertia about the new axis is $I_d = I_{sphere} + m\,d^2$. If we further make the new axis some distance Δ away from the center of mass of the sphere and locate it at the barycenter of the system for the orbiting object, we find that the orbital moment of inertia is as calculated in the previous discussions.

The parallel axis theorem is well known to mechanical engineers and is used in the design models for mechanical machinery with moving parts, such as motors or crank shafts or pistons or even spinning hard drives. The more complex moments of inertia require calculus to calculate, so one might assume that the concept of the spin moment and the parallel axis theorem were developed with the emergence of modern calculus over two hundred years ago. By that time, orbital equations were likely also being developed but the point-mass approach was employed, since it had about a hundred-year history of accurate use at that time.

At this point, we need to go back to the Lagrangian, L, and include the missing kinetic energy that exists for the rotating sphere. In the typical orbital models, there is only the radial and angular orbital motions supplying kinetic energies, and the single rotation per orbit is missing. The kinetic energy associated with the single rotation per orbit is $0.5 \, I \, \omega^2$, which when added into the Lagrangian, forms the complete Lagrangian used for calculating the orbital equations of motion that now includes the missing angular momentum. If a non-uniform or non-spherical object is tumbling or rotating at some new angular rate, we can see that this results in a fluctuating angular momentum relative to the direction of motion, and no such fluctuations would be predicted for the point-mass model. We need to include both the missed single-rotation and orbital angular momentum, but we also need to include the fluctuation of the angular momentum from the independent spin rate. While the mean angular momentum of the orbit is conserved, the instantaneous angular momentum fluctuates but is still conserved in that these fluctuating perturbations sum to zero over a protracted segment of an orbit or trajectory.

Using a few examples, we can show why the point-mass approach was considered a "perfectly" sound approach by finding the magnitudes and the impact of the "perturbation" on an orbit. The magnitudes of the correction terms can be seen by inputting the cylindrical dimensions for the Hubble Space Telescope spacecraft, which has dimensions $r_c = 2.1$m, $\Lambda = 13.2$m, and $\Delta \sim 6930$km using the mean radius of the Earth to find the barycenter of the orbit. With these values, and if the density of the Hubble is uniform, the perturbation on the orbital angular momenta is $\sim 2.25 \times 10^{-12}$ to $\sim 5 \times 10^{-15}$, depending on the dynamic orientation of the HST as it orbits the Earth. For a uniform spherical satellite with a diameter of 4.2m, the perturbation is $\sim 4 \times 10^{-14}$ and is constant with no variations. The perturbation for the sphere increases the total orbital angular momentum by a constant value.

For the Hubble pointing in a fixed direction, the scalloping only occurs with a period of the orbit, and it is doubtful such a perturbation could be detected except as a fixed perturbation of the estimated point-mass orbit that shows up after thousands of orbits. Any such variations or perturbations in the orbit would likely simply be attributed to the deviations of the Earth from a uniform sphere. And, in principle, we can locate and quantify the deviations in any object's shape and mass distributions by monitoring and measuring orbital deviations of any object orbiting another object.

For Earth-orbit crossing objects, we can make another estimate. Using

The First Anomaly: Orbital Angular Momentum

Newton's gravitational law, we can compare the gravitational attraction from the Earth on an object that is at Earth's orbital distance from the sun and see when the orbit about the sun becomes an orbit or trajectory influenced more by the Earth's gravity than by the sun's. It turns out that the two gravitational forces balance when the object is approximately half the distance of the moon away from Earth, though orbital perturbations become more important much further away. Thus, for a spherical object one kilometer in radius we calculate $p_{\theta\, tot} = p_{\theta\, cm}(1 + 0.4\, r_s^2/\Delta^2) \sim p_{\theta\, cm}(1 + 6 \times 10^{-12})$. When we consider the solar orbit alone, the perturbation is even smaller but increases in size as the influence of Earth's gravity becomes more pronounced in influencing the trajectory as the object approaches the Earth.

Consequently, any scalloping introduced into the trajectory by the Earth is unlikely to be measurable. The perturbation from the direct pull of Earth's gravity would be far more significant. Consequently, any influence on the trajectory by the angular momentum perturbation would only occur after the object strikes the Earth, since the barycenter of the orbit of such a small object is nearly at the center of the Earth.

Using other orbital parameters, we can find the magnitudes of the radial and angular scallop perturbations, which fall into the millimeter-per-second range of motions with concomitant distance variations of millimeters over the time span of the orientation fluctuations in the examples cited so far. This is discussed more in Appendix 1, but the upshot is that these are undetectable using current measurement systems.

We also discuss in Appendix 1 when the angular momentum perturbations may become important. The models for the orbital angular momenta show that when the size of the object becomes comparable to the distance of the object from the barycenter of the orbit, the perturbation becomes a large fraction of the total orbital angular momentum. But, more importantly, while the perturbation is a perturbation on the orbital angular momentum, it can look like something else, something we have all observed.

The moon orbits the Earth with one face locked to the Earth, which makes it appear as if the moon revolves once on its axis per orbit. Normally, there is no direct linkage between the rotation and the orbital angular momenta. (There are indirect linkages, though, through tidal effects.) However, using the exact definition for orbital angular momentum, we now have that the single rotation of the moon per orbit is really a component of the orbital angular momentum and, therefore, is fundamental to identifying the real orbital parameters of the moon.

Using a mean value for the value of Δ for the moon's orbit relative to the Earth-moon barycenter, we can find the magnitude of the perturbation for a uniform sphere. From these values, we find that the perturbation is $\sim 8 \times 10^{-6}$ of the center of mass value for the orbital angular momentum of the moon. What this means is that most of the values for what we know of the moon are off by a factor $\sim 8 \times 10^{-6}$, which includes the mass and the real orbital eccentricity versus mean values. Given a sufficiently long observation period, these small orbital perturbations accumulate into non-trivial and measurable orbital perturbations.

We can also see that there would also be a small solar angular momentum perturbation and that the moon's orbit is a complex Earth-sun-moon hybrid orbit. However, the moon's distance from the sun is vastly different from the moon's distance from the Earth. Therefore, there would be two vastly different magnitude perturbations with a period of one moon orbit about the Earth. To be sure, the moon's orbit has been a continual topic for academic study given the number of unexpected and unaccounted-for variations in the its orbit, and one source of these uncertainties would have to be the unaccounted for angular momentum and the variances that occur due to mass distribution and shape variations in the moon that cause a fluctuation in the orbital angular momentum in a single revolution.

The above arguments also hold for the planet Mercury. Putting in the values for Mercury, the orbital angular momentum has a perturbation $\sim 10^{-7}$ of the center of mass value. Thus, we have measured the eccentricity for Mercury's orbit and then attributed these orbital parameters to some point-mass angular momentum and planetary mass. Consequently, we have over-estimated the mass of Mercury by a factor $\sim 10^{-7}$.

A key point is that the dimensions of Mercury…or the moon or any space object…are really a component of determining the orbit of an object, whereas the spatial dependence is missing when using the point-mass approach. This same omission will appear later in other analyses, in which the dimension of an object and the physical extent of the mass distribution is important in understand the complete orbital dynamics of the object. In the case of Mercury, the rotation rate of Mercury and mass-density variations mean that the scalloping of Mercury's orbit and the associated motions and range variance should be measurable with current sensor technologies.

The First Anomaly: Orbital Angular Momentum

Where the angular momentum becomes important is when interacting objects come close together, such as for a collision or scattering event. When the approximation of one mass being much larger than the other no longer holds, the barycenter of the system is clearly not the same distance as used to define the mutual gravitational force between the objects. None the less, the general conclusions still hold. In astrophysics, once objects get close together, tidal effects can distort shapes and even cause objects to fracture. If two planetary masses of equal size come close to one another, such as when the primordial Earth spawned the moon, the barycenter of the orbits of the two colliding objects could be near the point of collision, in which case Δ can approach the radius of the objects in magnitude even as the mutual force using the point-mass model is defined by the distance $\sim 2\Delta$, which would significantly modify the orbital equations for the objects.

None the less, and for instance, if equal-sized spheres were just to collide, theoretically, the perturbation would become $\sim 29\%$ of the total angular momentum, which is a 40% increase in the total angular momentum and is too large to just be a perturbation. Of course, the objects are not generally in orbit if they are colliding, though one orbit could be decaying as the objects spiral toward a collision. If the objects are small enough, tidal forces may not be strong enough the tear either object apart and the objects might collide if their orbits had been perturbed, such as occurs for asteroids orbiting in the Mars-Jupiter corridor…the asteroid belt…or in the Kuiper belt. Of course, Saturn's rings were formed from tidal and collisional effects and the collision dynamics within a ring would have a small but anomalous angular momentum associated with the collision process that is currently not part of such models for the rings.

Note that for any orbiting object, the perturbation is always present and represents a "hidden" single rotation for each orbit. If the object's angular momentum is orientation dependent, then the common idea that rotational angular momentum has no impact on orbital parameters is false, though the effect may be too small to casually measure. However, in a near-collision scenario, the appearance of the collision would be for the two objects to look as if they were pivoting about their barycenter or contact point, and the actual radial motion of the objects as they pass one another would appear to momentarily "hesitate" while moving angularly. In appearance, the two objects appear to "roll" past the point of closest approach. Another way of describing the appearance might be to think of the objects as "sliding passed one another" as if the objects having some barrier that modifies the impact dynamics.

We discuss this phenomenon briefly in Appendix 1, where we show that when the rate of change in the angular rate is positive, the rate of change of the radial motion is negative.

With the increase in the angular rate comes a concomitant decrease in the radial rate, which means that the closing rate between the two objects decreases. Depending on their independent rotational rates, the effective spin of one object can appear to be increased and the other decreased…or both changed the same, though such changes are temporary. Even so, these effects could wreak havoc on the topology and structure of the objects during the encounter, if they are not within each other's Roche limits. Since the closing rate of the two objects is likely independent of their actual gravitational interaction…since they were presumably in separate orbits that brought the two objects into proximity…then the increase of angular momentum is a direct result of their mutual gravitational attraction as they come into proximity. Therefore, for some scenarios, the angular momentum increase would decrease the closing rate, possibly reducing the probability of the two objects impacting, though without detailed modeling, such collision dynamics could just as well be opposite as described above. A decrease in the closing rate means that the gravitational attraction acts over a longer period before the closest approach of the objects, which would clearly impact he interaction dynamics.

The scenarios are endless, though they all may have analogies with the early solar system. These scenarios are dynamically complex, because objects that come into proximity were likely in independent orbits about the sun before their mutual gravitational interact began to perturb these solar orbits. It is the perturbations that bring the objects into proximity, so that at some point their mutual interactions become strong enough to redefine their orbital dynamics.

The question might be, can we continual to claim that angular momentum is conserved during the entire scenario? At some point during the scenario, planetary mutual interactions will become a three-body problem and for these scenarios, we do not have a single central force. We could expand our perspective and look at the collision between galaxies. In these slow-motion collisions, we would see the granularity of the tidal forces in action as well as the effects of the angular momentum.

If we look at nuclear particles, we can have attractive and repulsive forces and the objects can have nearly equal dimensions or those more characteristic of an electron and a nuclear particle. Clearly, the scattering dynamics would be considerably modified when the angular momentum perturbation is included, and the attractive or repulsive nature of the forces would make the cross section of the collision more dependent on the modification to the angular momentum. Remember that the angular momentum is conserved because the mutual interaction is caused by a central force, which has no angular components as far as we know and as far as we currently model them. Consequently, while the total angular momentum is conserved, more of this momentum becomes a physical-rotational momentum during a collision.

However, if a particle, such a neutron has a magnetic moment, then the extra collisional "spin" could couple the angular motions in an unexpected way into the collision dynamics, possibly supplying un-noticed angular forces, in which case angular momentum might not be conserved, since the angular motion "steals" some energy and momentum from the radial motion. It is likely that our models for nuclear forces and for the spin of objects are subjected to unrecognized sources of error in the interpretation of the scattering cross sections as a function of the scattering angle. Our nuclear models are incomplete as is our understanding of the dynamics of the scattering process. Since we have no real physical description of quantum spin…it is just a number…now we have a way of introducing actual physical spin into the experiments, which should allow us to distinguish physical spin from quantum spin, if they are in fact different, since such quantum spin effects might be clearly manifest in collisions.

We could speculate over other scenarios in which the angular momentum plays a key role in defining the dynamics of the system. However, the goal of this chapter and all the gory supporting details were merely to show that our texts are not perfect and that analytical mechanics is not complete and that any physics or astrophysics which discusses orbital and collision dynamics is incomplete.

In the spirit of completeness, a survey of NASA web sites and various university web sites showed no awareness of the unknown perturbation. Furthermore, even advanced orbital mechanics and analytical or classical mechanics texts do not mention the orbital angular momentum perturbation, and all use a point-mass approach to the modeling. The classical point-mass models for orbital mechanics have been universally adopted. The perturbation should have been identified if known simply to dismiss it as inconsequential, which it is not in certain planetary physics and astrophysics.

The accuracy of the point-mass modeling for orbits has eliminated the needs for technologists to question any of the physics of orbital and celestial mechanics. More importantly, starting with secondary education through to graduate school, the idea of the validity of a point-mass approach is a core idea. It is no wonder that no one has questioned the validity of the concept. Yet, it is astonishing that there has been no thought given to why such a simple idea or concept as the center of mass seems to be so universally accurate. As the book progresses, the lack of such curiosity or lack of an associated but un-spoken skepticism will become more puzzling. Without skepticism there is no questioning. Yet, to be fair, from a practical perspective, the anomaly in the orbital angular momentum would never be observed. From an academic perspective, the anomaly should have been discovered over a hundred years ago.

Chapter 3—A Hidden Omission

Once the universal use of the concept of a point-mass model was successfully challenged, at issue is, where else have we used the center of mass and overlooked something important? Recognizing that we have gotten something as simple as orbital angular momentum wrong creates an incentive to see what else we may have gotten wrong. So, the next step is to see where the center of mass may have a hidden and lopsided importance.

Following the bread crumbs laid down by the falsification of the center of mass as the location of all of an object's mass as a universal analytical model, we have shown that angular momentum is a heuristic subjected to many levels of perturbation depending on any chosen scenario. The next step is to find where else the point-mass concept was incorporated into important physics and how we have erroneously used it. The investigations take us to written works by Newton, specifically his *Principia*.

The key issue is that Newton and his peers believed the point-mass concept was an absolute truth. They had no notion that the center-of-mass location for all of the object's mass was an approximation or heuristic for dynamic modeling. The only clue that perhaps certain concepts were incomplete came with Huygens' observation that the new period of a pendulum caused by adding an additional mass did not match the period predicted by using a center of mass to find the pendulum's swing-arm length. None the less, since the orbital dynamics as worked out by Newton did not consider conservation of energy directly and only considered angular momentum indirectly through the centrifugal force in defining an orbit, Newton was not forced to investigate other emerging concepts, which might have raise some questions relating to the center of mass as a completed notion within dynamics.

Just as Chandrasekhar revisited Newton's *Principia*, my attention was drawn to the law of gravity as established by Newton. It is a simply law, perhaps too simple. The predominant modeling in the *Principia* is from the perspective of the center of mass and point mass, and Newton used his dimensionless "corpuscles" as his mass elements. For the most part, the postulates and theorems and modeling are completely self-consistent within the *Principia*. It was only within the descriptions of how he was developing his gravity law for extended objects that there was a breakdown in the consistency. In retrospect, it should have been spotted long ago, but as Chandrasekhar pointed out, one must study the *Principia* line by line to find what it really says, and even then, the great man…and I mean that sincerely…missed what is likely the most important hiccup in our understanding of basic physics.

One barely must be a scientist or mathematician to spot where the problem lies within the *Principia* and with Newton's approach to finding the mutual gravitational interaction between two extended objects; because the issue relates to a simple flaw in mathematical logic. It is a matter of not doing what he was saying but thinking that he had done what he was saying. Blame for missing the logical flaw rests entirely with both the physicists and mathematicians of the ensuing three hundred years. As Newton might cynically opine, those shoulders of the giants we stand upon are not nearly as strong or as broad as we imagine or wish them to be. (Per Glashow, Newton was referring to Hooke and the idea of standing on Hooke's neck!) Yet, there is more to the omission than simply neglecting to look more closely at the works of the founder of analytical physics.

The analyses in Chapter 2 and in Appendix 1 motivated me to re-consider the origin and proof of the use of the center of mass of an object as being the mathematical location of all the mass of an object when performing dynamic modeling of the object. Newton employed the center of mass as a simplifying technique for his analyses, because it was believed by the natural philosophers of the time that it was universally true. The results of his efforts led to the subsequent extrapolation or inductive expansion of the idea that all the mass of *any* object can be considered concentrated at a point located at the center of mass of that object. Consequently, per Newton, since the mass of all objects is located at the center of mass as a point mass, the mutual force between two objects is only dependent on the magnitudes and separation of those point masses despite the shape, relative orientations, and mass or charge distributions within the two objects. This is a spectacularly broad claim and I was skeptical even though as a technologist I had used these physics successfully.

None the less, I researched how Newton had arrived at his hypothesis that the mass of any object could be considered concentrated at a single dimensionless point located at the center of mass of the object. I never found the exact origin to that belief other than references to Archimedes work in statics, though the concept could be older still, but it was a universally held belief that supported the observations during the renaissance. Then, as the renaissance natural philosophers and astronomers began to grapple with the dynamics of objects in motion, or at least for objects in orbit, the center of mass was a simple concept to employ and seemed to give excellent results. Consequently, Newton had no reason to question the use of the center of mass.

The proof supplied by Newton that the mass of an object could be considered concentrated at the physical center of mass as a dimensionless corpuscle was part of his description of what is now called Newton's law of gravity. This proof appears to have been first described in his *Principia Mathematica*, which was published in 1687 and written in Latin, which was apparently common in those years and limited readership until translations were produced. A search of other works of and about Newton's scientific efforts did not produce a definite contemporaneous reference to the center of mass concept except for the comment regarding Huygens's measurements of the period of a pendulum.

The *Principia* contains the first written description of the gravitational law that bears Newton's name. Hooke, a contemporary of Newton's, vigorously disputed the claim that Newton was first to describe the inverse square law of gravitation, whereas Newton claimed that gravitation and the law it obeyed was a matter of some learned discussions at the time among several other well-known scientists and that he, Newton, was simply the first to formalize a written description. Newton's claim has been supported as the correct interpretation of events. Again, Glashow's short paper highlights the contentious relationship between Hooke and Newton.

In reviewing the *Principia*, my attention was specifically focused on the pertinent description for the mutual interaction between extended objects as presented in Book 1, Section 12, Proposition 75, Theorem 35 of the *Principia* for the mutual attraction between two uniform spheres, which is reproduced in Fig. 3.1. The first literal translations of the *Principia* can be somewhat challenging to follow since they use colloquial English of the early 18th century. There have been multiple translations into more modern English even as recently as the early 21st century, plus there have been numerous analyses of the various propositions and proofs, yet the fundamental error was never noticed. Even though Newton was one of the originators of calculus, in the *Principia* the proofs were graphical and algebraic, which are the underpinnings to calculus but lack modern notations. Consequently, nobody tries to learn how to duplicate Newton's analyses, which may have contributed to missing the error.

The issue is that Newton made what would now be considered an unsupported and unproven supposition in his proof. His analyses were with respect to "corpuscles", which are infinitesimal points of matter. His proofs within Book 1 were primarily concerned with point masses, and it was only when he reached Proposition 75 that he introduced mutual attraction between spheres or extended bodies rather than between a sphere and a corpuscle or simply between two corpuscles.

Newton also introduced another more modern concept in Proposition 75, which was his Corollary 2, and it was both controversial and centuries ahead of its time. Corollary 2, shown in Fig. 3.1, is a rough paraphrase of part of what used to be called Mach's Principle relating to inertia, though one physicist chastised me for using an obsolete concept, which overstates the importance of Mach, who's importance has become much diminished since Einstein's time. Inertia was hypothesized to be a result of all matter in the universe interacting simultaneously with all other matter. In other words, each mass element in one object mutually interacts simultaneously with all the mass elements in the second object and vice versa. Additionally, mass does not shield mass and the mutual gravitational forces are additive.

> COR. 2. The case is the same when the attracted sphere does also attract. For the several points of the one attract the several points of the other with the same force with which they themselves are attracted by the others again; and therefore since in all attractions (by law 3.) the attracted and attracting point are both equally acted on, the force will be doubled by their mutual attractions, the proportions remaining.

Figure 3.1—Quote is from the translation of Newton's *Principia Mathematica* by Andrew Motte (1729): Book 1, Section 12, Proposition 75, Theorem 35, Corollary 2.

However, this is where our modern observational error has occurred. While we currently believe in the interpretation of Corollary 2 as being a legitimate statement of how two physical objects interact gravitationally, Newton proceeded to calculate the mutual interaction between two spheres by reducing one of the spheres to a point mass. And that is how we still do it.

In that regard, his result would appear to be legitimate and consistent with a sphere attracting a corpuscle and with the center of mass being a legitimate idea for locating all the mass of an object. All the point-mass approaches yield consistent results that reinforce the point-mass approach. This consistency lured us into accepting the results as an unequivocally true physics. None the less and unfortunately, Newton did not prove that reducing one sphere to a point mass was equivalent to the statement in Corollary 2.

It is also possible that Newton's focus on and commitment to point masses was a consequence of the near, if not the complete, impossibility of his being able to perform the modeling and calculations for two spheres rather than for a point and a sphere. Yet, Corollary 2 is what we currently believe to be an accurate and general statement of how the gravitational attraction between two objects occurs. Or, stated another way, if one were to ask physicists other than general relativists how they might graphically depict the gravitational interaction of two spheres, but not specifically to calculate that interaction, then they would paraphrase Corollary 2 and refer to Fig. 3.2c below as a graphical depiction of that paraphrase.

We can test whether the proposition that Newton stated in Corollary 2 that he wanted to solve, and which is explicitly what we now believe is a proper description, is in fact the same as the problem he did solve. We can then also test whether the problem that Newton did solve, which was to reduce one sphere to the size of a corpuscle, is equivalent to what he believed he was doing within the context of Corollary 2. In other words, we must prove that what he did is equivalent to what he thought he was doing for us to know that what he did produced was not a heuristic.

Figure 3.2 presents the essence of the issue. (The physical center of mass location is shown by the small dot in the center of each object.) Figure 3.2a represents the center of mass approach to modeling the attraction between two objects by considering all the mass concentrated at a single point at the center of mass of each object, and these point masses, called corpuscles by Newton, have no dimensions and are separated by the distance Δ. Figure 3.2b shows how Newton calculated the mutual force between two spherical masses, which is to make one of the objects a point mass and to find the mutual attraction between all points in the extended object with the mass concentrated at the center of the second extended object.

Figure 3.2c, on the other hand, is what Corollary 2 is telling us what Newton believed about the mutual interaction of two objects, spheres in his case. The centers of the two spheres are separated by the distance Δ but the mutual attraction is now from all points in one sphere with all points in the second sphere. Mathematically, we perform an integral, which is the mathematical way of summing all the mutual interactions combinations indicated by the solid line joining an arbitrary mass element in one object with an arbitrary mass element in the second object. We can use Mathematica® to help resolve the issue, though not without complexities discussed later in this chapter and in Appendix 2. I have not explored whether other computer programs might work as well as Mathematica®…or even better.

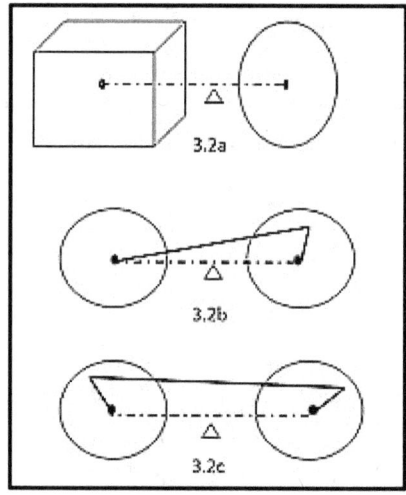

Figure 3.2—Three interpretations for the calculation of the mutual attraction between two extended objects, with the physical locations of the respective centers of mass separated by the distance Δ.

When we evaluate the integrals implied by Corollary 2, we find that Newton's simple law is a special case of the more general result. While Newton's gravity law is simple, it holds for a wide range of relative sizes of spheres and separation distances. None the less, there are many practical cases in which there are measured deviations, indicating that the modified and more general model is likely the correct model, but Newton's law as we use it is a heuristic, as good a heuristic as it is. In fact, it is so good we never suspected it might simply be an approximation. However, it is our very ignorance of the heuristic nature of Newton' gravitational law that aided the development and promotion of Einstein's formulation of the relativity theories, which is discussed in the second book of this series, *Einstein's Hidden Relativity*.

The detailed discussion of the integrals that we need to evaluate Newton's general gravitational model is relegated to Appendix 2. We will simply lay out the approach and some results in this part of the book. To start with, we need to review the traditional approach to this problem to identify any reasons that the exact approach was overlooked for three centuries.

The general problem of the mutual attraction between two objects can be described using the vector diagrams in Fig. 3.3. A similar description, and the only one I could locate, was presented in a widely-used text book entitled

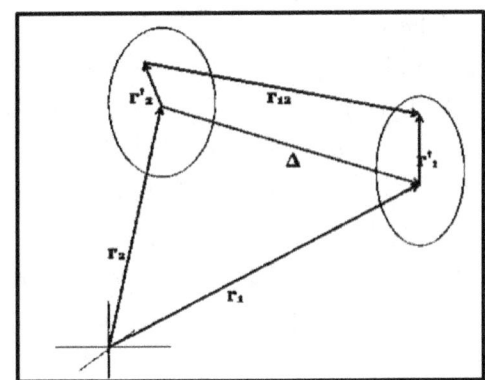
Figure 3.3a

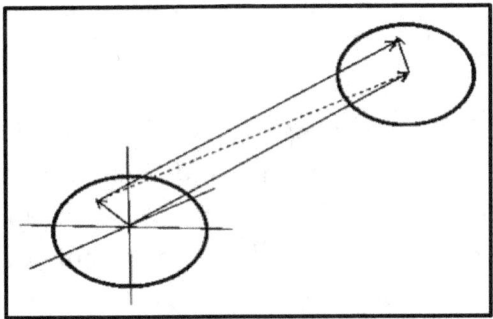
Figure 3.3b

Figure 3.3—Vector Models for the Mutual Interaction of Two Objects

Tensor Analysis by I. S. Sokolnikoff. All pictorial or vector descriptions follow the idea incorporated in Fig. 3.2b, which follows Newton's approach and which is being challenged here. Figures 3.2c and 3.3a follow from Corollary 2 in the *Principia*.

Figure 3.3 shows that, from some common origin, we draw vectors to the centers of mass of two objects, and the centers of mass are separated by a vector. We then draw a vector **r'** in each object to an arbitrary mass element in that object. Finally, we draw a vector $\mathbf{r}_{12}$ between the two mass elements in the different spheres. At this point we use Newton's inverse-range squared dependence to identify the force between the two point-mass elements, which are called corpuscles by Newton.

Sokolnikoff began by describing Fig. 3.3a, and then moved the origin of the coordinate system to the center of mass of one of the spheres, which is a good simplification, but he then collapsed the mass of the other sphere to a single point at the center of mass of that sphere, which he did for practical as well as historical reasons. The practical reason is that no one had tried to solve the more general problem and the tools of the time were likely not up to the task. Consequently, Sokolnikoff reverted to the simplification introduced by Newton without questioning that simplification. My questioning in class at the time was, why he had done that, and the reply was that that was the way it worked!

Once Sokolnikoff moved his coordinate origin to the center of one of the spheres, the vectors $\mathbf{r}_1$ and $\mathbf{r}_2$ disappear and we are only left with $\mathbf{r}_{12}$, Δ, $\mathbf{r}_1\mathbf{'}$ or $\mathbf{r}_2\mathbf{'}$ in Fig. 3.3b. Figure Fig. 3.3b without one of the primed values of **r** is the traditional but unproven simplification to the problem that Newton used. This geometry was examined in the early 1800s, using modern calculus and Legendre's mathematics to find a result that is still supplied in modern texts, which is the same problem Newton solved for the mutual attraction between a point mass and a spherical mass distribution.

The practical reason for keeping the simplification is that without making the point-mass assumption, the exact integrals, discussed next, are not analytical and it is only within the past few decades that it is in fact possible to perform the complete integration without the simplification. Plus, as stated before, there was no incentive to do more work than necessary, since the simplification gave a result that is so good no one suspected that it was a heuristic.

We can now review how our understanding of the mutual force has evolved. When we use the point-mass approximation, shown in Fig. 3.2a, we eliminate all volume integrals and simply put all the mass for each object at their respective centers. Thus, the mutual force is given as $F = G\, m_1 m_2 / |\Delta|^2$, where G is the universal gravitational constant and Δ is the distance between the two corpuscles. This law holds for point masses in which the mutual attraction varies as the square of the separation between the point masses. Consequently, since we can consider an extended object as made up of point mass elements or corpuscles, we have the relationship that the force between corpuscles…or centers of mass containing all of an objects' masses…can be found using $F = G\, m_1 m_2 / |\Delta|^2$. This will be referred to as the point-mass law.

In Fig. 3.2b, we require only one volume integral, because all the mass in the upper sphere is located at a single central point. The incremental mutual force in Fig. 3.2b, where the dotted line indicates that all the mass of the sphere is concentrated at the single point at its center of mass, is given by $dF = G\, m_1\, dm_2 / |\mathbf{r}_{12}|^2 = G\, m_1\, dm_2 / |\Delta - \mathbf{r'}_2|^2$ and where we are looking at the incremental force between the total mass m_1 located at the physical center of mass of the sphere and the incremental point mass dm_2 in the other sphere. To find all the mass in the second sphere, we must perform the integral over that sphere's volume in accordance with the model developed for Fig. 3.2b. Figure 3.2b is representative of how we still "prove" Newton's approach to his gravity law.

In Figs. 3.2c and 3.3b, however, the incremental mutual force is given by $dF = G\, dm_1\, dm_2 / |\Delta + \mathbf{r'}_1 - \mathbf{r'}_2|^2$. The masses dm_1 and dm_2 are the incremental point masses of the two objects, G is the universal gravitational constant, and Δ is the separation between the two centers of mass. Corollary 2 requires that the extra term in the second denominator be included. It only remains to determine the impact of this extra term on the mutual force between the total spherical masses. To do this we integrate over the volumes of the two spheres simultaneously and normalize the result to the same result from using the classical expression for the mutual force between two point masses. Newton failed to show that the results of the two approaches gave identical results, thereby simply stating and reinforcing the concept of the center of mass as the equivalent location for all an object's mass.

At this point, the details become more challenging, and the mathematical modeling efforts are relegated to Appendix 2. What we show is that the vector approach described above arrives at different mathematical descriptions of what Newton was trying to do. In fact, by explicitly incorporating the statement from Corollary 2, we have a different more generalized mathematical model than the one currently taught in all texts. However, to complete the analysis, we need to see whether the extra term in the model has any consequences.

So far, the descriptions that have been presented are either qualitative or symbolic. The issue is whether there are any mistakes in using Mathematica®, which is a fair question. The quick answer is that without a third party checking the actual calculations, we cannot know for sure. I did have expert third party confirmation although the interpretations of what was found, and the extent of the scenarios examined are uniquely mine at this point.

However, what should concern anyone when they review various published papers and reports, which were ostensibly peer reviewed, is whether it is possible to follow the derivation of any models that were used in an analysis. We also need to be assured that any analytical results derived from executing the mathematical processes defined by the models have been validated. None the less, often the numerical results are all that can be gleaned from an author's work, since the actual development of the models sometimes becomes hopelessly buried in the application of unfamiliar mathematics...if the document even provides a pathway for following the derivation.

In the simplest cases, one can look up the functions in various references to see how the various integrals or differential equations are evaluated. Where it gets tricky is when the models are not analytical and have to be evaluated numerically. Numerical methods are still being developed and, in some cases, such as for stability theory, for example, the calculation of a fundamental parameter called the Lyapunov exponent is fraught with uncertainty, requiring considerable investigation to ensure that the result is useful rather than convenient. Not only are such detailed checking not particularly well described, if at all, the subsequent numerical results are often supplied without any details on the programs used to make the calculations.

The issue is that many academically-employed computer programs are not widely available commercially if at all and are likely not used by the peer review people to double check that there has not been a computational error or hidden "assumptions." That this happens was demonstrated by a colleague whose model used in his dissertation research was an adaptation by a professor from yet another published model, but the results from using that adapted model for other data sets were not consistent with results reported elsewhere. It required seeking out the original model builder to identify a mistake that the professor had made and, ultimately, never acknowledged. Consequently, what is published is often likely far from validated and is used with some inherent risk, which is why when we cannot find an adequate second-source validation; we should be warry of using the results that have been published. The provenance or quality related to an author's affiliation is insufficient to ensure validity.

In fact, if you are reading this, it is incumbent upon you to be skeptical of my results until they are further validated. Consequently, my intent in this book is to supply the models that were used and the rationale for them. I used the Mathematica® User Groups Forums and a variety of text and user manuals to help me validate my methodologies and use of Mathematica®. Some of the forum members are Wolfram personnel, some are faculty who teach the use of Mathematica®, but most are simply user of Mathematica® in their professional roles. The learning curve to the use of Mathematica® must be followed carefully, but the power of the tool is enormous. I was also careful to avoid blindly accepting the results of using it, especially when the results were "very nice looking or provocative." There is more discussion along these lines in Appendix 2. None the less, the Mathematica® models used as the basis for all the modeling presented in this book will be provided upon request.

As stated previously, the generalized Newtonian gravity model needs to be validated against the common point-mass model. The approach to the validation of the generalized model was to perform a series of calculations, some of which results are presented graphically in the remainder of this chapter and the next chapter. The results of the general model were normalized point by point using the comparable results found using the point-mass model. A variety of scenarios were used to show that the results from using the general model present a granularity in the mutual force that is absent from using the point-mass model.

The comparisons using spheres were calculated numerically because there are no analytical forms for two-sphere integrals, contrary to the results from using the point-mass model. On the other hand, results of the modeling using spheres shown in this chapter, show that the gravitational interaction between uniform spheres supplies interactions that are closest to those of the point-mass model while showing that there is considerable small-scale variations in the data that, as stated before, do not impact any practical uses of the point-mass model. In the next chapter, we use other mass distributions for which the results of using the generalized model vary substantially from the results predicted using the point-mass model.

As shown in Fig. 3.4, we find the typical small variations in the mutual force between equal uniform spheres. Figure 3.4 shows the mutual force between two equal and uniform spheres as they are separated from contact to ten radii apart. It would logically be expected that at large separation distances of the objects, the results should approach the point-mass results and that is the case. The data plotted in Figs. 3.4 – 3.9 are for the mutual force ratio R, which is the general mutual force divided by the point-mass mutual force, which is then subtracted from 1.0. Thus, if the generalized force is larger than the point-mass force, $R > 1$ but the plotted value $1.0 - R < 0$. If the general force is equal to the point-mass force, $R = 1.0$ and the data point is zero, which lies on the x or separation axis.

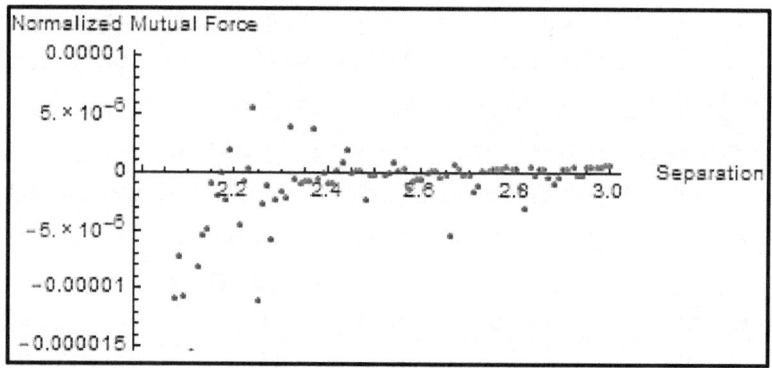

Figure 3.4—The data show the variance of the normalized mutual force away from the Newtonian point-mass force between two equal uniform spheres as a function of the separation between the centers of mass. The objects are in contact when the separation equals two. The results are typical in that at close range there is considerable though small variations in the mutual force away from the point-mass value. The variability diminishes with grater separation distance until the variations disappear and the results asymptotically approach the point-mass value.

However, the scattered data points when the spheres are near to one another raised concerns that the results may be indicative of noise in the calculated data. By this we mean that the numerical algorithms within

Mathematica® might cause such scattering about some mean value. To resolve the issue, many computational runs were made with variation in the precision and accuracy parameters and sample points in the coordinate systems used to define the geometric shapes. The conclusion was that computational noise could not be completely ruled out in certain scenarios, though in Fig. 3.4, the accuracy and precision were many orders larger than the computed variances. The computational noise refers to the various approximations employed by Mathematica® in carrying out the numerical calculations, and calculations using spherical coordinate descriptions often took far longer to carry out than the comparable computations using cylindrical coordinates.

On the other hand, Fig. 3.4 does not reveal all the salient features that are present in the data. Using a different presentation format and reduced range of separation distance, we can see in Fig. 3.5 that the mutual force has a significantly different behavior for the much smaller range in which the two uniform and equal spheres are just separated from contact. The contact mutual force is less than given by the point-mass model, and as the separation is increased the mutual force approaches the point-mass value and then becomes larger than the point-mass value by a small amount. For larger separations, the mutual force asymptotically approaches the point-mass value as the separation is further increased.

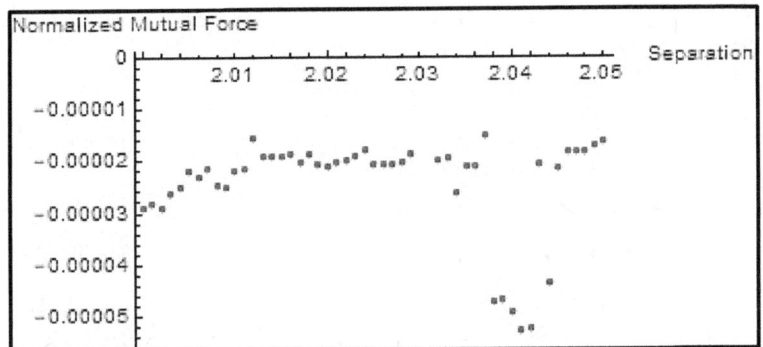

Figure 3.5—The data are presented for the small separation distances used to calculate the data in Fig. 3.4 near the contact range in which Fig. 3.4 does not show any results. The objects are in contact when d equals two. The results are typical in that at close ranges there is considerable though small variations in the mutual force away from the point-mass value.

In the next chapter, we perform some calculations on spherical shells, and results are not too different than for solid uniform spheres, with similar data variances as shown in Figs. 3.4 and 3.5. However, when objects such as cylinders are interacting, the mutual interaction shows even larger though smooth deviations from the point-mass results, and the smoothness persist for nearly all separation distance, including near proximity of the objects. Again, at sufficient separations distances, the results do approach the point-mass results, as would be expected, which is shown more clearly in Fig. 3.6.

In Fig. 3.6, we have calculated the mutual interaction between a small sphere and a large sphere as the small sphere is separated from contact out to 10 radii of the larger sphere. The scaling was that the larger sphere had a unit radius and the small sphere had a radius of 10^{-7} of the larger sphere. Scaled to Earth dimensions, the small sphere would be ~64 cm in radius. In Fig. 3.6, for d > 4, the variances in the data were on the order of the precision and accuracy selected for performing the calculations. For lower

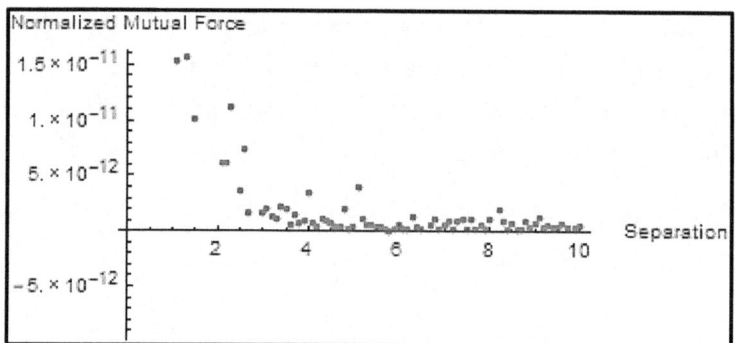

Figure 3.6—The data points represent the normalized mutual force on a 64 cm radius uniform sphere relative to a uniform Earth-sized sphere as the small object is translated from contact with the Earth's surface at d = 1.0000001 to an altitude of ~ 63,780 kilometers at d = 10. The results for d > 4 are approaching unity asymptotically and the variances are in the range of ~ 10^{-12} of the point-mass values.

precision and accuracy, chosen to speed up the computations, the data would show little-to-no deviations from the point-mass value. The computation time for the models often took many days at high precision, accuracy, and working precision settings, so that many parametric evaluations were required to establish some pattern for the deviations of the results from the point-mass values.

Figure 3.7 shows the results of using the same objects as used in Fig. 3.6 calculated at still higher precision over a much shorter separation range. The separation d = 1.00001 corresponds to an altitude above the Earth's surface of ~ 64 meters. The variations in mutual force away from the point-mass model persist over considerable separation distances but are qualitatively nano-variations. In fact, using the approach described previously for plotting the data points, the mutual force at contact is slightly less than the point-mass value by the amount shown on the vertical axis of Fig. 3.7.

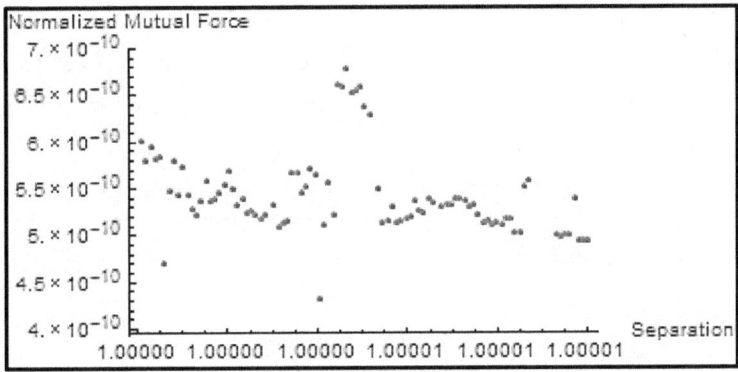

Figure 3.7—The data were calculated using the same objects as used to find the data shown in Fig. 3.6. The accuracy and precision used here are higher, though, and the separation ranges are smaller and represent the data very close to the vertical axis in Fig. 3.6.

One final comparison is shown in Figs. 3.8 and 3.9 In these two figures, we have placed two uniform spheres in contact and varied the radius of one over a wide range. Therefore, rather than finding how the mutual force varies with separation, we find how the mutual force varies as size ratio is varied. Two figures were used to show how the granularity of the data changes over a wide-range of comparative sizes for two spheres. For very small radii, the variations shown in Fig. 3.8 are on the order of microvariations in the relative magnitude of the mutual force and remain consistent over many orders of magnitude radius variation. For larger radii ratios, as shown in Fig. 3.9, the variability becomes

A Hidden Omission

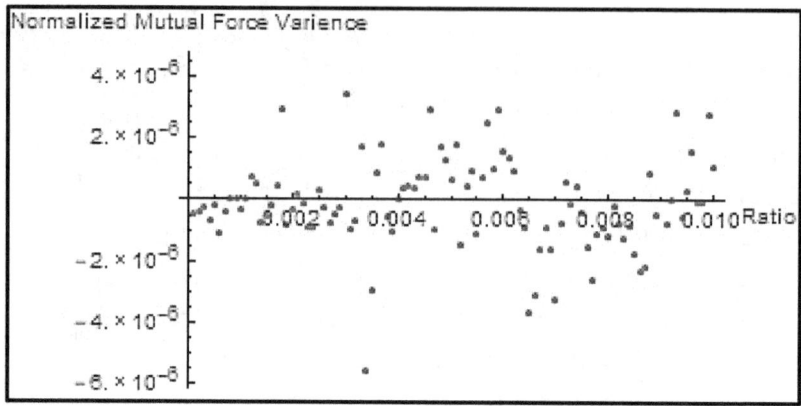

Figure 3.8—The variance of the normalized mutual force between two uniform spheres in contact as the radius of one is varied. The results are typical in that for spheres in contact there are considerable though small variations in the mutual force as the radius of one sphere increases, and the variability increases with increased sphere size ratio.

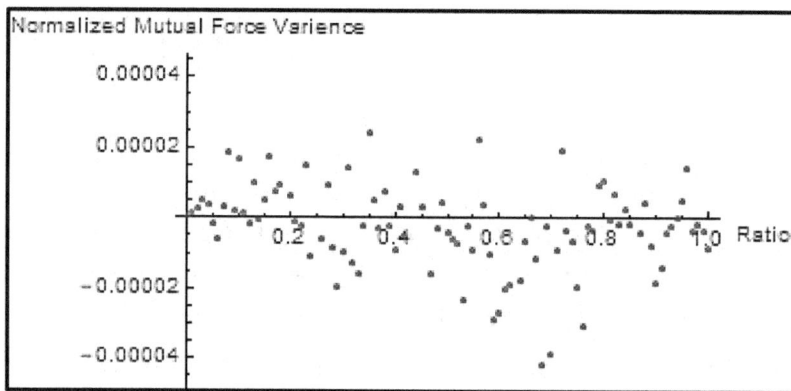

Figure 3.9—The data, as that shown in Fig. 3.8, show the variance in the normalized mutual force between two uniform spheres in contact as the radius of one is varied, though in this figure the size ratio of the spheres is much greater than in Fig. 3.8.

measurably larger as the two spheres approach equal size. When the two spheres are of equal size, the variance is consistent with the data shown in Fig. 3.5 when the two unit spheres are in contact with a separation of the centers of mass of two units. Even though the variances in the data in Fig. 3.9

are on the order of 10^{-4} of the point-mass values, there do not seem to be any practical consequences to the variability.

On the other hand, the small variations shown in the figures could impact our understanding of the standard mass used to define the world standard for the kilogram. The sensitivity of any given measurement would have to be sufficiently great to detect such small variations with altitude relative to sea level. Since spheres are not compared in finding the standard mass, the data shown here is only indicative of the possibility of there being some slight impact on measurements for the standard mass. The results shown in the next chapter for cylinders might be more realistic in terms of expected variances.

Also, given the relative sizes of the smaller object and depending on how rapidly it is changing position relative to the center of mass of a larger object to which the small object in interacting, some small vibration or stress patterns might emerge as the objects separate. This would be more relevant for larger objects rather than for a smaller object separating from a larger object. Given the nano-variances, though, mass density variations within the larger object would likely cause more variances in the mutual attraction as the objects change their relative positions.

The accuracy and precision used in producing the figures were orders of magnitude larger than the variations, indicating that the data are likely not noise but show a real effect, small as it is. The data are also indicative of the ways in which the general gravitational law produces variations that are absent when using the point-mass law. In a general way, the mean values in the data sets are nearly given by the point-mass law.

One conclusion from the above results is that size does matter in finding the mutual force between objects. Non-uniform spheres and objects show even greater and wider variations in the mutual force between objects, just as we found that the angular momentum depends on the mass distribution and relative orientations between objects. The gross effect of the variations is to supply a non-inverse-square effective mutual force, and from work by Newton, such forces cause orbits to precess. In extrapolating these observations, it is likely that instability or chaos in orbits is exacerbated by these variances in the mutual force. Whether the perturbations on the mutual force by mass-distribution or "out-of-round" variances are larger or smaller than the non-inverse-square perturbations will need to be assessed. Perhaps as with the angular momentum, such variances only show up in specific circumstances and perhaps not at all for practical technologies.

A Hidden Omission

One place in which scientific work could be impacted by the application of the general Newtonian gravity model is in measurements of the universal gravitational constant, which is discussed further in Chapter 4. We also mentioned how we arrive at the standard kilogram and how that standard may not be as fixed as we assume. However, in this chapter we will remark on the consequences to astronomy relating to the mutual forces between the Earth and the moon and between the sun and Mercury. The detailed modeling is reported in Appendix 2, but here we will simply mention what the models indicate using a limited number of calculated scenarios.

The planetary data include ranges of separations that may never have occurred or may occur at some far distant time, which indicate that current orbits fall within a range of separations that only show very small variations away from the point-mass values but do show that there are small gradients in the data. The data were calculated assuming the objects were uniform objects of the same density, where the density only supplies a multiplicative constant that is cancelled by normalizing the results using the point-mass mutual force under the same circumstances. Presumably the data plots spanning a larger range of separation show the historical mutual force before the objects settled into their current orbits. Such data could also help identify the orbital stability within the early solar system.

In performing the calculations, the separation distance of the moon from the Earth spanned a range of $56 < d < 63$ in terms of the radius of the Earth, and the radius ratio was 0.273. For Mercury, relative to the sun, the radius ratio was 0.0034 and the separation distance in terms of the sun's radius was $66 < d < 100$. The normalized mutual force was calculated as a table of values and the resulting tabular data were plotted as shown in the figures.

For Mercury, the mutual force with the sun is shown in Fig. 3.10, and the range of the mutual force only varied by $\sim 10^{-10}$ from the point-mass values over the current range of the orbit of Mercury. Given the small variances of the normalized mutual force away from the point-mass mutual force, it is easy to see why the point-mass model was never doubted. As small as this variation from the point-mass law may be, over protracted periods of times…years or centuries…the orbit of Mercury would precess and the perihelion of the elliptical orbit would advance by a small amount per orbit. The calculation of this precession was one of putative crowning glories of Einstein's general relativity, which found a variation from the point-mass law, but then the general Newtonian law also shows sch a precession. However,

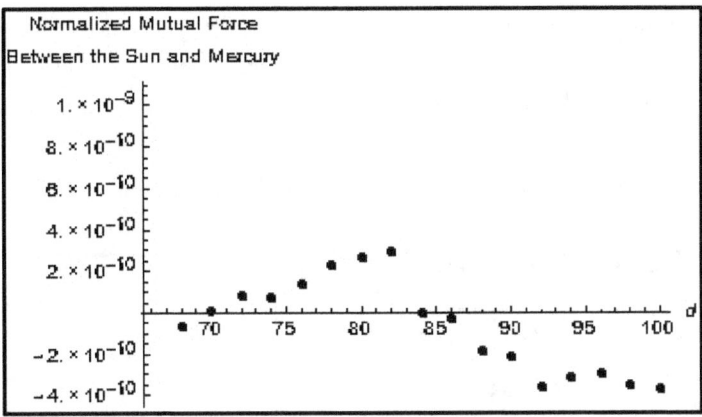

Figure 3.10—Normalized mutual force between the sun and Mercury for the separation range of 66 to 100 solar radii. The general variance from the point-mass mutual force in the current orbit of Mercury is on the order of 10^{-10}, which is not an obvious deviation of the mutual force away from the point-mass value but suffices when integrated over a century's number of orbits to cause a measurable advance in he perihelion of Mercury's orbit.

both objects are taken as uniform spheres and we have ignored the impact of the sun being an oblate spheroid, which may produce further variances in the mutual force that essentially established Mercury's current orbit and precession.

We also calculated the mutual force between the Earth and the moon, ignoring the fact that the solar gravitational affect is comparable in magnitude to the direct Earth-moon interaction. From a size ratio and distance perspective relative to the sun, based on the results of the analysis for Mercury, the variances in the mutual force between the Earth or the moon and the sun are at least many orders of magnitude smaller than the variances for Mercury and the sun. To first order, we can compare the Earth-moon mutual force with that for Mercury and the sun to determine if the moon is experiencing greater or lesser mutual interaction variances than Mercury.

Figures 3.11 and 3.12 show the normalized mutual force between the Earth and moon for separations of between three and a hundred Earth radii. Figure 3.12 shows the data calculated for the current separation distance. For the current orbit of the moon, the discontinuous variances of the normalized mutual force relative to the point-mass mutual force in the current orbit is $\sim 10^{-8}$, whereas the variance for the much smaller separation distance such as during the formation of the moon is ~10^{-6}. Given that the Roche limit for the moon is ~ 3 Earth radii, the mutual force for the separation of less than three Earth radii could never occur. The precision and accuracy of the calculations is higher than the variances, indicating that these variations away from the point-mass values are likely real.

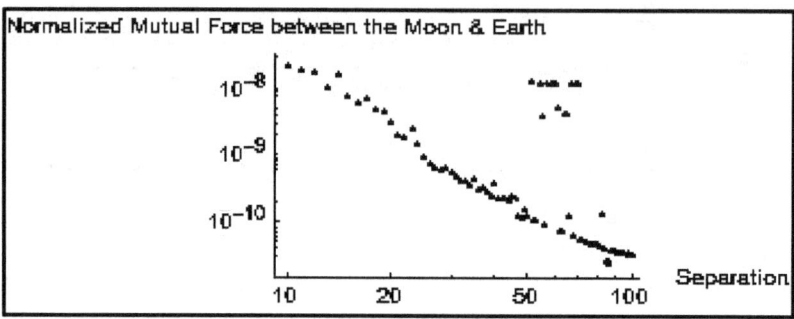

Figure 3.11—Normalized mutual force between the Earth and the moon for the separation range of from 2 to 100 Earth radii. The data for the current separation range is shown in Fig. 3.12.

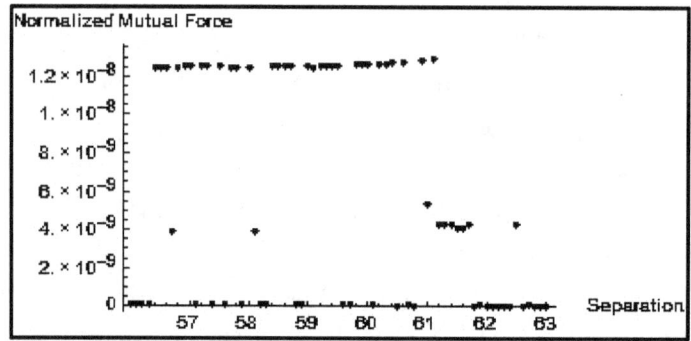

Figure 3.12—Normalized mutual force between the Earth and the moon calculated over the current orbital range of 56 to 63 Earth radii. The calculations were the same as for Fig. 3.11 but limited to the current orbital range. The discontinuous variance over the current range is shown in both Fig. 3.11 and 3.12 and are much larger than the precision and accuracies of the calculations.

All the figures only compare how the variance in the mutual forces depends on the relative sizes and separation distance between spherical objects. At the orbital distances of the moon from the Earth, Figs. 3.11 and 3.12 indicate, in addition to the unusual behavior for the current moon orbital range, that there are variances in the mutual-force curve that could produce an orbital precession from the deviation of the mutual force from an inverse-square law. One may speculate that the evolution of the Earth-moon separation since the formation of the moon has experienced significantly different dynamic behavior than we suspect, though it is not clear what this may mean. All we can surmise now is that the moon experiences unrecognized forces during an orbit that greatly complicate our understanding of lunar orbital dynamics. While the above results are interesting, it is not clear what these results mean in a practical sense.

What all the results do indicate is that the mutual forces found using the general Newtonian gravitational model are more granular and variable than the simple point-mass models would indicate. While these results may have an impact on academic research, from a technological perspective, it is not as clear what may be impacted by the newly identified granularities in the mutual forces. One could speculate that space objects may experience both internal and orbital perturbations that have long-term consequences on both orbital dynamics and structural or internal integrity of an object. Certainly, multi-orbit stability will be impacted, perhaps significantly.

In the next chapter, we look at some other shapes and combinations of shapes, which reveal that uniform spheres are the objects that are closest to those that obey the point-mass gravitational model, at least for objects that are externally separated from one another. The calculated mutual forces on small mass elements or objects within a mass distribution show smooth variation even with high precision calculations, but the shapes of the curves are totally unpredicted by the point-mass model.

Yet there are some ranges of sizes and separations that indicate that we may be viewing quasi-steady state conditions both within the solar system and the galaxy rather than the more deterministic conditions that we have historically attributed to the orbits of the planets. That is, the current orbits of objects may be the result of billions of years of chaotic motion that has resulted in the orbits settling into their current apparently or even putatively stable configurations.

What we can say is that the point-mass mutual interaction models are idealizations and that real interactions contain variances and data trends that are totally missing when using the point-mass model for mutual forces. We may conclude that the general law is revealing the heuristic nature of Newton's point-mass or point-charge interaction models. Practically speak, however, the variances in the mutual force are likely limited to better understanding of orbital dynamics for satellites and spacecraft, though we cannot eliminate the possibility that electron energy distributions within solid may reveal unknown characteristics when the general model is used in the quantum calculations. And, as noted, we need to refine what we mean by the standard mass, which is critical in supplying a baseline for certain measurements across the world's societies.

Missing from all the prior analysis is the impact of real shape and density variations on the mutual interactions. We have considered some relative size variations among uniform spheres, but that is all. The results in the next chapter indicate that once shapes and densities are perturbed away from uniform densities for perfect spheres, the mutual force variances become significant. Therefore, the oblate spheroidal shape of rotating stellar objects may have a substantial impact on the local planetary system's dynamics and mass distributions. In fact, these force variances may ultimately be shown to force planetary systems into the ecliptics in which they exist as quasi-stable orbiting systems. Any rogue objects orbiting out of the ecliptic could be inherently subjected to destabilizing forces that lead to their ejection from the system.

A practical consequence to these variances would be to better understand how spacecraft orbits within the solar system or around planetary objects might become perturbed over time. The inverse is also true in that the orbital perturbations can be more precisely identified with mass and shape variances within the primary object about which the spacecraft is orbiting. We already do this, but without understanding the variations caused by the non-inverse square gravitational forces, we are not making the correct attributions to the sources of orbital perturbations.

In addition, the point-mass gravitational or electronic interaction models have for two hundred years employed an approach called the multipole expansion. The multipole expansion is a particular series expansion of the function $1/|\mathbf{r'}_2 - \mathbf{r'}_1| = 1/\delta r$, where $\mathbf{r'}_1$ and $\mathbf{r'}_2$ are the vector locations of a point within each object relative to the objects' centers of mass and as shown in Fig. 3.3a. The simple multipole given as $1/\delta r$ is the results of using the point mass as the form for the interactions of objects.

While this form looks correct, it only connects two points together and not two volumes, as discussed previously. When the requirements of Corollary 2 are included, we must use the form $1/|\Delta + \mathbf{r'}_2 - \mathbf{r'}_1| = 1/|\Delta + \delta\mathbf{r}|$, where Δ is the vector distance between the two centers of mass. This form only reduces to the point-mass form when Δ is much larger or smaller than $\delta\mathbf{r}$. Consequently, the multipole expansion is also a heuristic that quickly breaks down as $|\Delta|$ approaches $|\delta\mathbf{r}|$.

In addition, deviations in the shape or mass (charge) distributions within an object are currently accounted for by using an approach that "perturbs" the result by using the point-mass models to model mutual interaction. These perturbation models are used to account for shape and distribution deviations away from those for a perfect sphere with a uniform distribution. The basis of perturbation theory is the multipole expansion, and so the perturbation model is also a heuristic. However, we have shown that the gravitational force is also "perturbed" relative to the point-mass models for perfectly uniform spheres, in which the multipole approach to perturbations would show no perturbations. The conclusion is that we must know the relative sizes, shapes, orientations, mass distributions, and separation distance of the interacting objects to accurately calculate the actual mutual interaction.

And, as a blow to the simple point-mass and multipole mathematics, we have shown that it is the mutual size and orientations of the objects that determine the actual form for the mutual interaction and for the fields. Therefore, a single object does not have a field associated with it. At a minimum, only pairs of objects give rise to something called a field and the magnitude of the field is more complex than predicted by the point-mass or point-charge models. This is a blow to the field theories that were introduced during the development of post-classical physics at the turn of the 20th century. While the heuristic may work for large separations of field-producing sources, interactions are often short ranged, and the field models become much different when sources are in near proximity.

Even for quantum theory, an electron or a charge is not necessarily a point entity except at infinity. For any other interactions, such as in solids or during collisions, we need to understand the charge distributions and sizes of objects that are interacting. Since quantum theory is supposedly the highest accuracy approach to modeling interactions, we have that quantum theory based on point charges and point masses is itself a heuristic and not as inherently accurate as we have presumed, especially for near-field or close-proximity interactions.

So, what have we accomplished so far? We have identified all the above unexpected caveats from questioning how Newton arrived at the point-mass approach to gravity! The point-mass heuristic is the underpinnings to nearly a hundred years of advanced modern physics and the foundations are merely heuristics! Our approach to the physics is that we have observed perturbations and deviations from our theories, and from these observations we have adjusted them by inventing new physics to correct these models and mathematics, yet we have not corrected the underlying physics and mathematics.

We may also surmise that the early focus on potential fields is an abstraction that no longer has any meaning. The early adoption of the potential likely occurred because these forms simplified the production of analytical models and results. Again, these are idealizations with limited value and, perhaps, have played a role in our lack of curiosity relating to possible deviations of physics from our early simple models and understanding. Since the more complex and exact mutual gravity results require numerical integration, the potential models also require numerical integration from which we must make further numerical calculations to find the mutual forces. Consequently, the use and value of the potential functions will become marginalized, since the simple analytical forms of the potentials are simply heuristics that may or may not work well enough for practical purposes.

On the other hand, even the current perturbation theories use quantitative data to adjust these forms to fit the observations. Consequently, even though the generalized Newtonian gravitation law shows that there are "perturbations" away from the point-mass law, we use adjustments associated with mass distributions that are clearly not uniform and objects that are oblate spheroid, miscellaneous attraction form other celestial objects and tides that distort mass distribution to identify what are empirical models that seems to meet the requirements of observations.

The more exact theoretical models do not necessarily improve either predictions or descriptions of mutual interactions. We may better understand that the center of mass as a heuristic or that the orbital angular momentum also a heuristic, but it is not clear that we have achieved anything more than alert ourselves that we have closed the book on classical gravity far too soon.

Chapter 4—Further Discussions on Newton's Gravity

To reiterate the conclusions from the last chapter, if the analytical results hold for the new general Newtonian gravity, there will clearly be astrophysical and planetary physics ramifications in terms of orbital stabilities and collision dynamics. However, practical issues would seem to be limited to unwanted or unexplained orbital perturbations for spacecraft or asteroids as well as some challenges in identifying what we really mean by the standard kilogram.

There may, however, be further practical consequences to our lack of knowledge regarding the mutual gravitational forces between objects. One consequence may be that we are overestimating the stability of orbits of any kind. The stability of orbits can be assessed by calculating the Lyapunov exponent for any dynamic scenario. With a more precise way of calculating the mutual gravitational force between objects, the dynamics will be perturbed away from the inverse-square models. Consequently, we should expect there to be more instability in orbits over various epochs. Also, certain anomalies that are observed in such space maneuvers as gravitational assists may now be understood to be a result of not employing the proper models for the trajectories and speeds, which currently use the point-mass concept.

Since the deviations from Newtonian gravity are very gradual and since most measurements are based on local calibrations, it is hard to identify where the deviations from Newtonian gravity might have an impact on non-academic Earth-based measurements and systems. However, there may be consequences to our methods of measuring certain fundamental constants or, as mentioned previously, in creating standards. As this chapter progresses and we look at other shapes rather than spheres, we will see that there are both size and orientation issues in establishing a standard for mass. Furthermore, altitude and location matter and more than just for the centrifugal forces arising from the rotation of the Earth. In addition, the measurement of the universal gravitational constant will require a new review of the sizes, shapes, and locations of the test masses being using to find this constant, which we will also revisit later in this chapter.

Another consequence of stability theory is that the orbits of stars in the galaxy may be decidedly non-Newtonian both inside and outside of the central core. In fact, outside of the galactic core, the orbital speed of the stars in the galactic disk is practically constant and nearly independent of the distance from the center of the galaxy. There are only a handful of theories to explain this non-Newtonian behavior. As this chapter progresses, it will be pointed out that part of the issue is that a test mass embedded within some mass distributions will have decidedly non-inverse-square net forces on it, which may locally appear to be Newtonian but, in the aggregate, are not. This holds for stellar objects embedded within the total mass distribution of the galaxy.

One of the more prevalent and controversial gravitational and dynamical theories for the galactic rotation, which is about thirty years old, is what is called the MOND (Modified Newtonian Dynamics) theory, which requires among other things that there must be uniformly distributed "Dark Matter" in the galaxy that dominates the galaxy's mass. The MOND theory is really a heuristic model that purports to explain the observed galactic stellar orbital speeds. In a theme repeated over and over in this book, such heuristics are not the physics but are only empirical models that fit or seem to fit the observed data.

The controversy with MOND is that some astrophysicists want the heuristic to be the physics. The heuristic can be an indicator of where and how to address the underlying physics. However, it still requires some underlying physics to justify its form. Hence, Dark Matter was hypothesized to account for the galaxy's rotation, so that the MOND heuristic might have a physical underpinning. As we will show at the end of this chapter, a lack of understanding of Newtonian gravity is likely the underlying reason for the galaxy's "peculiar" rotation characteristics.

The new general Newtonian gravity might supply at least a part of a possible explanation for the MOND observations. What most people do not realize is how complex the shape of the galaxy is, with the stars outside of the central bulge region being in a ~1-2 kilo-ly (lightyear) thick band that is made up of multiple spiral arms with large gaps between the arms. The bulge itself seems to have a peanut-like shape in certain directions (or a bar, since the galaxy is a barred galaxy with the bar embedded in a large oblate spheroidal central core). The core is roughly 10 kilo-lys in diameter…roughly an oblate spheroid whose shape flares at the galactic plane as the spheroidal distribution flows or flairs into the planar ecliptic distribution. As will be shown later in this chapter, the mutual forces can be far from those used to model the solar system, and as with the solar system, the mutual forces are not entirely what we expect from just considering the inverse-square or point-mass mutual forces. Even if the central galactic bulge is taken as a large slightly oblate sphere, the consequential mutual attractions of the stars external to the bulge, while decidedly not following a simply inverse-square-force relationship, might yield a velocity curve required by MOND. But then, maybe it might not.

Another interesting issue arises when we re-do the models for the gravitational attraction on a test mass that is within a solid or hollow volume of mass. If there are radial variations in density, a spherical geometry is more appropriate than the cylindrical geometry used in the earlier models. What these various approaches do show is that the results obtained from our modeling can be far from the results obtained by staying slavishly with the inverse-square force law.

Using our current inverse-square Newtonian models, we find that a spherical shell of mass or charge has no "gravitational field" anywhere within that shell. The reason is that the coincidence of an inverse-square law calculation for the force within a spherical mass shell turns out to be a zero at all locations within the shell. This occurs because, within a spherical volume, the attraction of mass in any solid angle is countered by an equal and opposite attraction in the mirror image of the solid angle. The coincidence of the inverse-square force law and the square law for defining solid angles mean that in any given direction taken from any point within a spherical shell, the forces always cancel. Consequently, the inverse-square force shows that the shell of mass or charge outside of the radial location of the test object supplies no force on the test object and, therefore, all forces cancel. This leaves the core within the radial location of the mass or charge element as supplying any net forces on the embedded mass element.

On the other hand, while the gravitational force within a hollow sphere may be zero, there is a constant potential field. Since the force is found by taking the negative gradient of the potential at any point within the volume of the sphere, the force field may be zero but the potential field can be a finite constant. When we generically refer to fields, we are being purposely ambiguous. And since we only know a field exists because of the action of the field on a test mass or test charge, we are left with the idea that we cannot attribute a magnitude and distribution of a fields without knowledge of the objects (at a minimum two objects) producing the field, which includes the test object.

However, while there is an unlimited number of scenarios that could be modeled, and dozens were in preparation for the discussions in this book, some scenarios are more revealing than others. For instance, shells or hollow spheres show, for external test masses, about the same amplitude variations as for solid spheres for the scenarios in which the mutual force is calculated as the external separation distances between these shells are varied. In general, the results are like those results found from the calculated mutual forces for solid spheres. The results support the contention that spheres do in fact behave as quasi-point masses with some perturbations on the calculated results. As we will show later, however, non-spherical shapes can show significant differences from the point-mass model.

More interesting were the calculations for the mutual force on a test mass within both a hollow sphere and a solid sphere. Using the point-mass model, for a solid uniform sphere, the mutual force on a test mass goes to zero linearly with the radial location of the test mass. In this case, the general law also goes to zero as the radial location of the test mass goes to zero. Therefore, when we use the point-mass law to normalize the calculations for the mutual force on a test mass, we find a finite value for this force as the test mass location approaches the center of the solid sphere.

On the other hand, if the test mass were located within a hollow sphere, the normalized model is indeterminate in the hollow volume, since the point-mass mutual force is identically zero everywhere within the hollow portion of the sphere. Consequently, if there is a force found using the general model, we could not normalize that force using the corresponding point-mass value.

As a result of the above observation, the interior mutual forces for certain models were arbitrarily normalized to the mutual force at the surface of a sphere using a test mass with a radius ration of 10^{-6} to 1, which for an Earth-sized planet would be a test mass ~6.4 m in radius. Also, all mutual force curves do go to zero at the centers, which would be expected by symmetry within a sphere, which is not necessarily true for non-symmetrical distributions. Additiionally, it is notable that, over many orders of magnitude in radius, the size of the small test mass did not materially affect the normalized mutual forces. However, as shown in Chapter 3, the finite size of the test mass does not quite reduce to the point-mass result when the generalized model is used as the test mass is reduced to a point mass.

We can graphically show the results of the above descriptions and can compare the results from using the using the inverse-square law by itself and as a normalization factor, which shows the deviation of the point-mass results from those obtained using the generalized law. Consider using the same normalization as described above in which we have a point test mass of mass m on the surface of a uniform sphere of radius R. We then place the point test mass within and at some distance r from the center of the solid uniform sphere. The only force on the test mass at this interior location is from the mass within the core with a radius of r. When we calculate this force using the point-mass form of Newton's gravitational law and normalize it to the surface mutual force, we find that the normalized mutual force on the point mass is $F = r/R$. This shows that using the point-mass and the inverse-square law, the normalized mutual force is a straight line having value unity at $r = R$ and zero at $r = 0$. The results of using the point-mass model are shown in Fig. 4.1. Keep this in mind as we compare this result with the results of using the generalized law. Figure 4.2 shows the consequence of using the generalize law in the same scenario depicted in Fig. 4.1. In both these figures, the results are normalized to the mutual force on a point mass at the same location. Consequently, since the mutual force curve using the point mass model is $F = r/R$, the curve in Fig. 4.1 goes to unity at $r = R$.

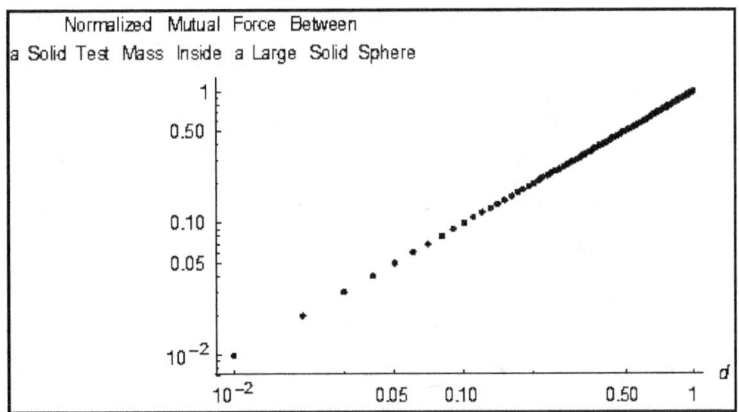

Figure 4.1—Mutual force on a small test mass with a radius 10^{-8} of the outside radius of a solid sphere using the *point-mass* Newtonian model for the mutual force. The normalization is with respect to the value at r = R.

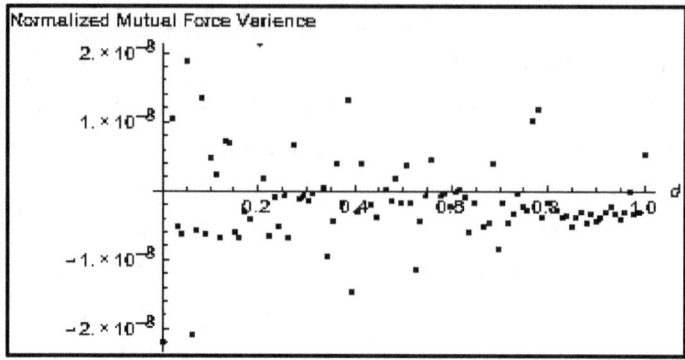

Figure 4.2—Mutual force on a small test mass with a radius 10^{-8} of the outside radius of a solid sphere using the *generalized* Newtonian model for the mutual force. The normalization is the same data as shown in Fig. 4.1 for each data point, so we are showing how the generalized mutual force compares point by point with the point-mass law.

Figures are often plotted in log-log scaling because of the amplitude ranges in the calculated results. Therefore, we have shown what the linear model for the interior mutual force looks like in log-log presentations. As noted before, keep this in mind as we show the results of several models for mutual forces inside hollow spheres and within other non-spherical shapes.

Figure 4.2 shows that the generalized Newtonian law predicts essentially the same net mutual force as the point-mass models, even though the general law accounts for the attraction of the mass in the shell above the location of the test mass, unlike the point-mass model in which this same attraction is identically zero. A key point is that the internal forces increase as the test-mass location increases toward the surface of the distribution, unlike the external forces that decrease with separation distances. In general, the internal mutual forces on small test masses exhibit variations with range similar to that for the external mutual force shown in Figs. 3.6 and 3.7. While these variations exist for spherical object, this is not the case for the mutual force for non-spherical objects, which we show later in this chapter for cylindrical and disk-shaped objects.

Next, we looked at the results for finding the mutual force on a test mass within a hollow sphere, including finding the mutual force when the test object is also embedded within the solid shell of material. As the radial location of the test mass is reduced from the surface toward the hollow-core interface, the resulting mutual forces go to zero as the mass remaining between the hollow core interface and the location of the test mass goes to zero. The variance between using the point-mass law and the general law are shown in Fig. 4.3. These variances are on the same order as for the solid sphere. Consequently, the general mutual force model does not produce any significant deviations from the same results calculated using the point-mass model, though the small order variances may have some significance for certain scenarios. From a practical perspective and for idealized uniform spheres, any consequences of using the general law may be restricted to academic research.

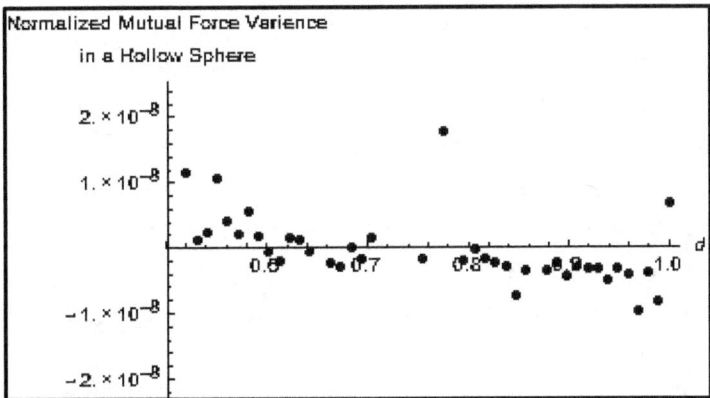

Figure 4.3—Mutual force on a small test mass with a radius 10^{-8} of the outside radius of the hollow sphere. The inner radius of the hollow sphere is 50% of the outer radius, which is also the outer edge of the hollow core. The test mass is moved from the inner surface out to contact with the outer surface and is zero mutual force within the hollow region.

There are two aspects of the mutual-force modeling results that bear further discussion. First, over a range of many orders of magnitude for the test-mass radius, the curves are similar with similar features with only some slight variations in the magnitude of the features. Therefore, the size of the test mass is not critical in the models if the test mass is very small compared to the larger mass. Second, the curves are reasonably smooth and not subjected to scrutiny as statistical or computational effects. This latter point is because the accuracy and precision of the calculations are much greater than the variations, indicating that the variations are real.

We have a final set of figures that indicates that the mutual forces between objects are sensitive to the shapes and the relative orientations of certain objects. We looked at the mutual gravitational interactions between uniform right cylinders of various lengths as their separations are varied from contact to some distance apart, which were modeled using the both the length and separations specified as multiples of the radii of the objects.

The easiest objects to model were uniform right cylinders aligned end to end. In the following data, the radii are by convention all. Mutual forces between shells or internal test masses were not modeled, though they could have been. The choice was made not to do such modeling and to move on and leave such details to the future. As with the internal models for the spheres, we would need to model scenarios that are relevant to some goal.

Let's look first at some uniform solid cylinders that are oriented end to end. The specific cylinder sizes are identified in the caption to each figure. The separations are given in terms of the distance between the centers of mass of the two cylinders given as multiples of the radius. Lengths are also given as multiples of the radius. Of particular interest is the interaction between cylinders with lengths equal to the diameter of the cylinders. These plots indicate the sensitivity of the mutual forces to exaggerated deviations from spherical shapes.

Figures 4.4 - 4.6 show the mutual force between equal uniform cylinders aligned end to end. These cylinders are moved from contact to some separation distances, and the figures indicate how quickly...or slowly...the

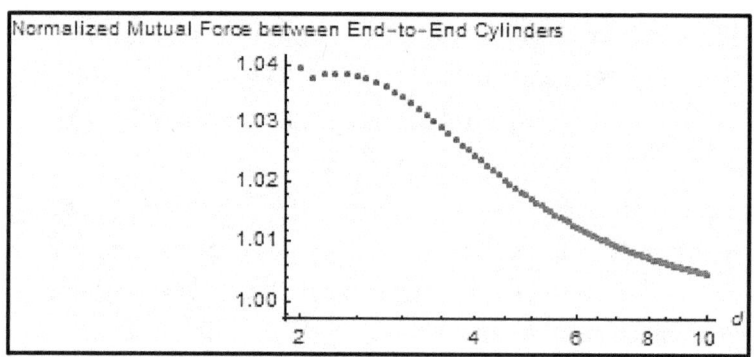

Figure 4.4—The mutual force is shown between two equal uniform cylinders of length equal to the diameter positioned end to end, which are separated from contact (d = 2). The small discontinuity at d = 2 is shown in more detail in Fig. 4.5.

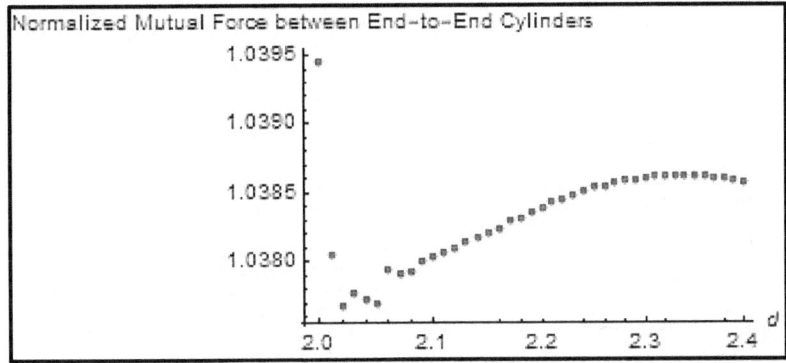

Figure 4.5—The mutual force between the cylinders used in Fig. 4.4 is shown in more detail for the small discontinuity in the region 2.0 ≤ d ≤ 2.4.

mutual forces approach the point-mass results with larger separation distances. All curves are normalized to the point-mass value at the same separations, so that the curves are showing the deviation of the actual mutual forces away from the point-mass values. Except for the contact and proximity locations, the data are much smoother than for spheres. However, the deviations from the point-mass model are also greater than for spheres, reinforcing the fact that each scenario must be individually modelled. There is no quick way to find the mutual force for end-to-end cylinders. In fact, Fig. 4.6 for longer cylinders shows that deviations from the point-mass model can become significant.

We can also compare the end-to-end scenarios with a side-by-side scenario. Again, we used uniform and equal solid cylinders to find the mutual force. In Fig. 4.7 we show the side-by-side results for the same two cylinders modeled in Figs. 4.4 and 4.5 where the lengths equal to their diameters. These side-by-side cylinders are moved from contact (d = 2) to ten times the radius apart, and the deviation of the forces away from the point-mass values is more striking than for the end-to-end geometries. For long narrow cylinders, the cylinders must be moved almost 50 radii apart for the resulting mutual force to become nearly the point-mass value. Figure 4.8 shows this behavior for cylinders with a length equal to six times their radius. This degree of variance from the point-mass model is also in stark contrast to the less extreme interactions for end-to-end cylinders.

We can outline the modeling for a scenario of interest to astronomers, which is only partially developed in this book. If we review the earlier description of the galaxy as a barred spiral galaxy, there are four key features,

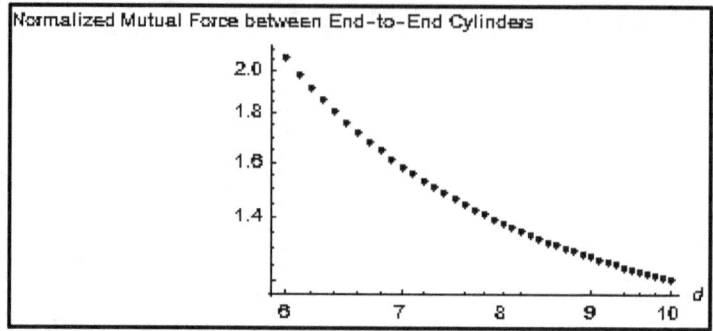

Figure 4.6—The mutual force between two equal uniform cylinders of length equal to three times the radius and oriented end to end is shown from contact (d = 6) out to six times the cylinders' radii. There is no discontinuity for the near-contact proximity as shown in Figs. 4.4 and 4.5 for shorter cylinders. The contact mutual force is also significantly larger than the point-mass value.

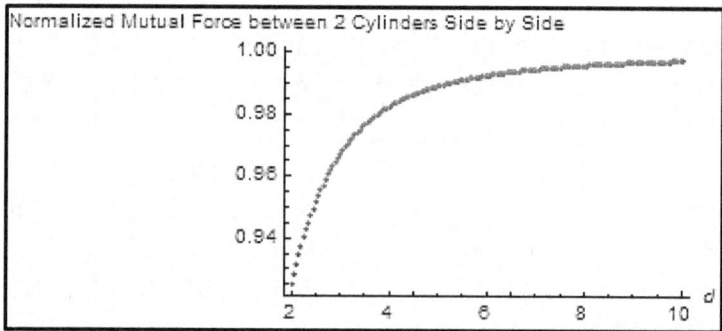

Figure 4.7—The mutual force is shown between two equal uniform cylinders of length equal to their diameters and oriented side by side and separated from contact (d = 2) out to ten times the radius. The contact mutual force is significantly smaller than the point-mass value.

which relate the shape and mass distributions within the galaxy. First, the bar is like a cylinder with a particular mass distribution, which is embedded in a core of stars. Second the core is also roughly an oblate spheroidal shape that tapers or flares into the galactic disk, and this core has its own mass distribution. Third, the disk is really a tapering disk that itself fans out at the outer rim of the galaxy to a thin layer of stars, and, again, the disk has its own

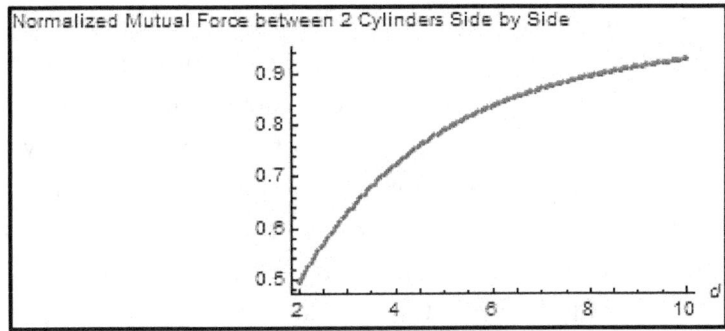

Figure 4.8—The mutual force is shown between two equal uniform cylinders of length equal to six times their radii, oriented side by side and separated from contact (d = 2) out to ten times the radius. The contact mutual force is significantly smaller than the end-to-end value.

mass distribution. And forth, the galactic disk is comprised of two major spiral arms with several large inter-spiral gaps that are mostly empty of visible stellar objects.

What the above description means is that any model for the gravitational force on a test mass…a stellar object…located anywhere within the galaxy will require the summation of the net forces from several different distributions on the test mass, all of which have their own unique contributions to the net force on the object. We have already looked briefly at the mutual force between two perfect and uniform right-circular cylinders. This model can be easily adjusted to form one of the cylinders into a small sphere and the other into a thin disk. We used this model to look at the mutual force on a small spherical test mass that lays on the perpendicular bisector of a cylinder and in the center of the galactic disk or ecliptic, which is the mutual force on a test mass embedded within a thin disk.

In attempting to scale the cylinder to the size of a galactic bar, it became obvious that our knowledge of the size, shape, and mass distributions within the Milky Way galaxy is "rough" at best. It appears that the galactic bar is ~20,000 ly (light-years) long by ~ 10,000 ly in diameter with the sun very roughly ~ 25,000 ly distant at an angle of ~ 45 deg off the center line of the bar. For the model, these dimensions are used to find the ratios of dimensions relative to the radius of the bar, which determines the functional form for the mutual force. The mutual force shown in Fig. 4.9 is for the test mass (sun) lying along the perpendicular bisector of the galactic bar at ~ 25,000 ly. We can see that the distance of the stellar mass must be > ~ 5 bar radii distant for the mutual force to be asymptotically close to the point-mass mutual force. In so far as the bar contains some fraction of the total mass of the galaxy, we can see that the bar's mass yields a distinct non-point-mass force on stellar objects within the galaxy. Not shown in the figure is that the mutual force varies almost linearly from the edge of the bar out to 10,000 ly with a minimum at the bar's outer surface of ~ 0.45 times the point-mass value. The mutual force on the sun is ~ 0.94 the point-mass value.

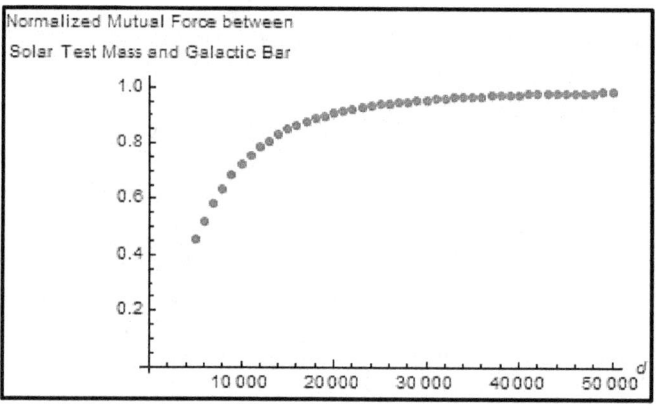

Figure 4.9—Mutual force between a uniform right-circular cylinder and a small spherical mass located on the perpendicular bisector of the cylinder. The cylinder is scaled to the size of the Milky Way's galactic bar and the small spherical mass is scaled to a solar-sized object.

The galactic core was taken as a well-defined spherical shape having a radius ~ 5,000 ly. The core's mass would be minus the bar's mass and volume. Ultimately, the modeling becomes an exercise in super-positioning the mutual forces of the intersecting bar and core on a distant object, which is left to some graduate student somewhere. It is likely that the effort will be more complex that the simple description given above, since removing a certain distribution will change the effective net mutual force because the shape of the distribution has changed. This is not an issue with the point-mass mutual-force model.

Finally, we can make a further estimate on the additive mutual forces on stellar objects embedded within the galactic disk, where the disk can be very roughly modeled as a 1,000 ly thick mass of stellar objects that is a washer-shaped ring with an empty inner disk of radius ~ 5,000 ly. The outer ring would have radius ~50,000 - 90,000 ly, though it could be more or less. The disk clearly tapers to a thin outer edge and the stellar density is far from uniform. Estimates are that ~10% of the galactic mass may be gaseous. However, the mass distribution in the galaxy are made using the point-mass model plus the presence of so-called dark matter. Consequently, the magnitudes of the additive components of the forces on a stellar object within the disk are, at best, guesses.

There is considerable ambiguity in the dimensions, mass distributions, and mass fractions of the galactic mass associated with any portion of the galaxy. The bulge, which excludes the bar volume, is estimated to contain 10^{10} times the mass of the sun. If we scale the diameter of the sun to the diameter of the bulge, we find that the bulge diameter is roughly 10^{11} times the diameter of the sun. Consequently, we can crudely estimate that the bulge or core density is essentially the same as the mass density of the sun. There are other estimates that the bulge mass is approximately one-sixth the mass of the galactic disk. The rest of the mass of the galaxy is made up of halo stars, gases, and the putative dark matter. Consequently, in finding the orbital forces on the sun in attempting to account for the suns orbit within the galaxy, the use of the point-mass models merely confounds the various estimates of the galactic mass distributions, including the more-or-less homogenous distribution of the dark matter. These estimates are a boot strapping exercise using the point-mass mutual force model. In other words, currently such modeling is an exercise in futility.

It is, however, instructive to see how an embedded stellar object on the ecliptic is bound within a disk of distributed mass. We have seen how the mutual force varies on a test mass within a uniform spherical mass distribution, so we will perform the same modeling on a test mass embedded in a thin uniform disk. We have a model for two cylinders that are oriented side by side, which we can modify by adjusting the cylinder limit for one object to that of a thin disk and that of the other object to be a sphere, which is essentially the model for the stellar object on the perpendicular bisector of the galactic bar.

In finding the mutual force on a test mass embedded within a uniform disk, we are ignoring the apparent granularity of all the discrete stellar masses within the galactic disk or even the galactic core. However, there is surprising consistency in terms of the distributed mass densities within of the galactic disk and core with regard to numbers of objects and their masses when compared to the solar mass. At a granular level where stellar objects are close and discrete, we would get local variances in the mutual forces. However, in the aggregate, we can use the sun as a mass element within a uniform disk to find the mutual force on an object such as the sun.

One issue is that there is no useful normalizing term, since the force on an object embedded within a uniform disk is not, surprisingly, a well worked out model, though as we will show, it is straightforward to find the force on test mass, using the sun as a test mass. The mutual force at the center of the disk on the central plane of the disk would be zero. Also, the mutual force from the disk material at radial ranges greater than the range of the test mass from the center does not go to zero as for a sphere. Given the poor understanding of the mutual force within a uniform disk using the point-mass model, we can simply normalize the distribution using the mutual force on the object at the outer rim of the disk. This is arbitrary but does allow the deviation of the mutual force away from a simple inverse-square distribution to be more clearly shown.

Figure 4.10 shows the normalized mutual force on a test mass (the sun) embedded on the ecliptic plane within a disk with a central hole. In this case the hole in the disk is the diameter of the galactic core, which is 10,000 ly. The calculations using the new model are taken beyond the galaxy's radius to show how the mutual force might impact an extragalactic object that lies in the plane of the disk. In Fig. 4.10 we used the mutual force on a point-stellar object at the outer rim of the galactic disk to normalize the plot. We also used a positive value of the mutual force to represent the mutual attraction of the masses toward their centers of mass. Figure 4.11 is an enlargement of the mutual force in and near the open central area of the disk and shows that the force from the disk on the test mass goes to zero at the edge of the hole at ~5000 ly from the center of the galaxy. The data are normalized so that the mutual force goes to unity at the outer rim of the disk.

The results shown in Figs. 4.10 and 4.11 are without a doubt nonintuitive and unpredictable. The normalized mutual force using the point-mass model on the disk distribution to find the force on a test mass at 50,000 ly is significantly higher than given using a point-mass force model which considers all the disk's mass to be concentrated at the center of the disk. The comparison shows that the normalized mutual force for

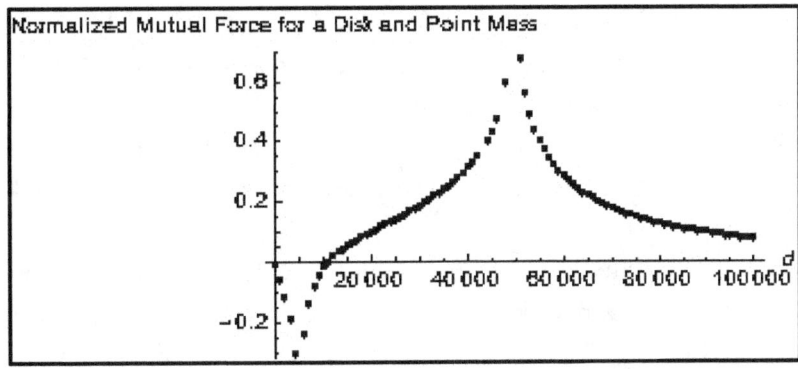

Figure 4.10—Mutual force between a uniform sphere (stellar object) scaled to 7 x 10^{-8} light year (the sun's radius) when the object is on the central slice of the ecliptic within the galactic disk. The data are calculated for ranges from the center to the rim of the disk and embedded in the "horizontal" slice at the center of the disk. A positive force indicates the attraction of the centers of masses of the two distributions. The mutual force is shown from the center of the galaxy to 50,000 ly out past the galactic rim. The galactic disk is taken as uniform with a central 5000 ly radius hole in the center where the galactic core would reside. The data are normalized to unit value at the rim of the disk.

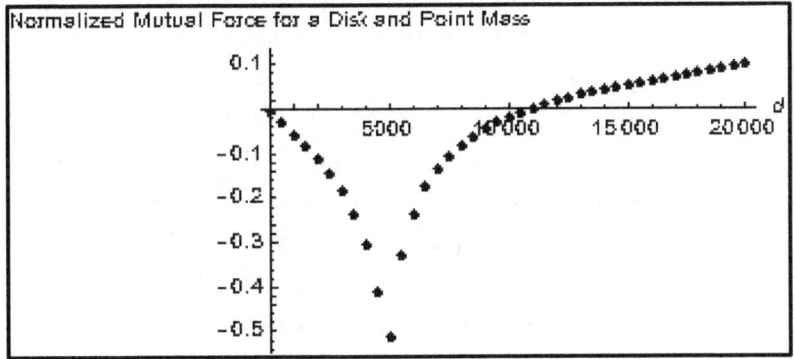

Figure 4.11—This figure shows the mutual force from the region in Fig. 4.10 in which the mutual force becomes repulsive, indicating that the stellar mass is being attracted toward the outer portion of the galactic disk more strongly than it is being attracted toward the center of mass of the galactic disk.

the distribution at 50,000 ly is ~ 3.6 times the point-mass disk model, and that at 1 Mly the normalized mutual force has fallen to 1.003 times the point-mass mutual force, where the normalization was for the test-mass mutual force at the galaxy's outer rim. We can see that in the plane of the galactic disk, at galactic distances, the mutual force has fallen back to the point-mass value.

However, the mutual force within the disk near and within what would be an empty volume of space in which the core resides is much more surprising. This region is plotted in Fig. 4.11. We see that as the stellar mass distance to the center of the galactic "ring" begins to drop to zero magnitude at ~ 10,000 ly from the center and then becomes negative as the object is closer to the galactic center of mass, indicating an apparent repulsive force on the stellar object by the disk. The peak negative or repulsive value occurs at the inner edge of the ring at 5000 ly from the center of the ring. What is physically happening is that the outer ring closest to the object is attracting the mass more strongly than the more distance disk mass that is angularly more distant from the location of the stellar mass.

Consequently, the cumulative mutual force on an embedded stellar object everywhere within the galactic mass distribution requires that we quantify and sum the mutual forces on an object from the galactic bar, from the core, and from the disk. Once the embedded mass is off-axis relative to the galactic bar and the galactic disk, the models would become more complex but in principle amenable to numerical solutions. It is little wonder that the classical approach to understanding the rotation of the galaxy has failed, requiring the invention of new physics such as MOND and dark matter. However, if the dark matter distributions were known, then if it exists, its effects can be modeled and compared to the results found using only the visible matter distributions. By clinging to the point-mass model, we completely misrepresent the actual galactic gravitational forces on any stellar object either within or external to the galaxy.

There are many other mutual-force scenarios, most of which are of academic interest. For the cylinders, however, the modeling becomes relevant to the measurements for the universal gravitational constant. What the models indicate is that the historical inability to achieve high accuracy in the determination of the universal gravitational constant is likely a result of the mutual forces within the measurement apparatus being far from point-mass mutual forces. The figures in this chapter also indicate that the mutual interactions of arbitrarily shaped objects in planetary scenarios are also often far from point-mass mutual interactions.

We could extend our speculations and observations into other cosmological scenarios that include the interactions of galaxies. And, to continue the scenarios, the consequence of the mutual forces within the galaxy being far from point-mass mutual forces may explain why theorists have been confounded as to why the galaxy is shaped the way it is as well as why our galaxy has the unusual uniform rotations that contradicts the results from using the point-mass models.

With a better understanding of the mutual interaction between objects embedded within a mass distribution, we may also begin to answer other questions that have emerged in the past several decades, which are fundamental questions that have, up to now, had no possibilities of being answered. For instance, none of the current laboratory experiments checks for whether or not the universal gravitational constant is really a constant. We have observations of the early universe that seem to indicate that perhaps G is not actually a constant, but those observations may be an artifact of our lack of knowledge about the deviations of the local mutual forces between galaxies away from the inverse-square law. Whether G is affected by the mass near where it is being measured or whether G has evolved in some fashion over time are topics ripe for further study. More than likely the closer the observations are to the "birth" of the cosmos, the more likely that the actual mutual interacts are like those for a test mass within a spherical mass distribution. Therefore, our models for the dynamics we expect are using the wrong mutual forces and accelerations. Still, from a practical perspective, the accuracy of measuring G has little or no impact on non-academic technology.

However, there are other factors associated with the mutual forces that can also impact dynamical systems. We still do not know if mass shields mass and we do not know how fast the mutual forces propagate. Newton himself was troubled by the apparent propagation rates having to be instantaneous if objects were to maintain stable orbits. Speculation on the propagation of the gravitational force has attracted some fringe science. On the other hand, one of the luminaries of 20th century physics has commented on how arbitrary Einstein's choice of the gravitation force propagation speed as the speed of light really was. If we have really detected gravitational waves, then the synchronization of the arrival of these waves with the visual observation of cosmic events…and even with the simultaneous detection of neutrinos from these types of events…would seem to corroborate the propagation speed as being nearly the same as the speed of light.

The results shown in this chapter clearly indicate that the generalized Newtonian gravity model predicts mutual interactions that are not consistent with the point-mass models. Shapes and orientations of interacting mass distributions are clearly important in determining mutual forces and mutual potentials. The models also show that spherical distributions show the closest results to those obtained using the point-mass models. None the less, if we use a criterion of which model impacts technology, we find that the generalized model only impacts what can be characterized as academic physics. The inability to fully explain observations by using the point-mass models has given rise to what may be characterized as pseudo-science used to explain what the point-mass models have not explained or predicted.

Chapter 5—General Conclusion on Newton's Gravity

This book describes a hybrid research effort in that the history of the various heuristics associated with the point-mass or point-charge models were investigated to understand why these were not recognized as heuristics and why no further attempts have been made to ensure we fully understood Newtonian physics. To be sure, the totality of what we call Newtonian physics has not been discussed here, and the physics of Newton's dynamic laws of motion are the subjects for a later book and a new invention. However, Newtonian physics relies on certain assumptions such as the point-mass and inertial formalisms. By uncovering the various scenarios in which the point-mass assumptions lead to wrong conclusions, we have identified inductive reasoning as likely a faulty approach to certain science.

The formal definitions of inductive, deductive, and abductive reasoning do not play a significant role in physics except for those who chose to dwell on these formal philosophical ideas and approaches. We have processes that are by consensus good approaches to problem solving. A problem is simply something we don't have an answer for. Starting with Newton, we called our educated guesses "hypotheses" and Newton pondered how he arrived at these hypotheses or educated guesses and concluded that he did it via inductive reasoning. In reality, Newton seemed to be looking for ways of simply generalizing his findings as he extrapolated into more complex scenarios. How Newton or any of us operate intellectually seems to be an unstructured combination of these more formal processes. What is it when we have a hunch and make a successful guess about something? Lucky? The blind pig finding a truffle? Or is it when we observe something and concluded that the explanations for what we have observed are wrong, non-existent, incomplete, or stupid and that we can do better?

That there is likely no such thing as a single scientific principle is a sentiment that apparently echoes that of Nobel Laureate Steven Weinberg. Popper called it problem solving. The issue is how we determine what problem deserves our attention or, worse, we simply don't see problems where there are problems. We have even, in this new era of correct-speak, eschewed the word problem and replaced it with the word challenge. The challenges are both observing and naming the problem and then attempting to solve the problem. Popper would have loved our modern double-speak. Einstein opined that once the correct question is asked, often the path toward a solution reveals itself…of that I am not so sure. For myself, a healthy dose of skepticism seems to be the sauce for the goose.

The identification of the vast number of heuristics in fundamental physics was not a result of a hypothesis or even a guess or hunch, though it was motivated after forty years by a single instance of performing a calculation and getting an answer that was not consistent with what I had been taught. The further motivation after forty years was that I was not able to sell an idea for a technology that was a novel exploitation of known technologies and physics. It was the physics that was stumping the technologist with whom I spoke. Consequently, unwilling to take no for an answer, I set out to make sure that the underpinnings to what was being described was universally understood to be true. I was propelled toward doing the research for this book by internalizing a quote attributed to Richard Feynman that, and I paraphrase, if an idea cannot be explained simply then perhaps that idea is not understood at all. So, as the King said to Alice, "…start at the beginning…"

The first small beatings of the butterfly wings have, consequently, identified errors within a variety of sub-disciplines within physics and astronomy. In the case of the orbital angular momentum, we have the beating of two butterfly wings. One was the orbital angular momentum itself and the other is the concept of the center of mass. Thus, we have transitioned from the first wing beats to the beating of a small flock of butterflies. However, a third wing beat has occurred in that we have evidence that the absolute perfection of certain aspects of analytical mechanics is in fact unsupported. This third wing beat is introducing a new perturbation into what has become an unrecognized and deleterious rigidity within the thought process of academic physicists toward physical phenomena and modeling. Whether the issue is not asking the right questions or refusing to ask these questions or simply staying with asking the questions they can answer rather than asking the hard question remains to be sorted out.

Whether the physics described by the angular-momentum perturbations will grow into something substantial remains to be seen. From my perspective as a technologist, I see the consequences as purely academic at this point. None the less, the orbital angular-momentum perturbation is not a consequence of anything Newton did, right or wrong, but is rather a consequence of post-Newton era physicists not questioning the center-of-mass approach to orbital mechanics or more broadly to non-linear dynamics. Further, for hundreds of years, we have not paid attention to what is in our text books. We did not extrapolate the use of the parallel axis theorem for non-mechanical interactions or, for that matter, for any orbital dynamics with or without a physical tether or constraint forcing motion along an arc.

Observational evidence that there may be an issue with the center of mass as a concept simply did not arise historically, since few observations exist to challenge the idea. Historically, only Huygens's observation of the pendulum period perturbation would seem to have indicated an issue. Consequently, there was no obvious incentive to look further into the concept, even in the twentieth century when quantum mechanics was introducing evidence that classical mechanics failed. The point-mass heuristic was simply blindly accepted and never questioned. Why was that? In a discipline that prides itself on attention to details, missing the orbital angular momentum perturbation was a surprising oversight. Such an oversight may say something about the susceptibility to and reliance on heuristics by physicists. At first blush, heuristics persist because they can allow physicists to "think" faster about physics.

On the other hand, the complexity of the moon's orbit should have stimulated a re-evaluation of orbital kinematics and the various sources of orbital perturbations. The demonstrated lack of curiosity has been a collective hallmark of physicists as they have closed the book on further research into various physics and have chosen not to re-investigate ideas that seemed to work to see if these physics are really heuristics. Such an attitude of accepting the work of the historical "giants in the field" without re-visiting their work seems to be an attitude unique to physics. The brilliant French physicist Leon Brillouin made a similar observation in his book *Relativity Reexamined*.

As part of the push back on the angular momentum perturbation, several professors who have taught advanced classical mechanics were on the one hand astonished that I would suggest that there were oversights and errors within the texts with which they taught. On the other hand, they dismissed the finding once they "got it" stating that it must be known to people who work in orbital mechanics. They showed no further interest. Whether their skepticism was that I or any non-academic in general could make such a discovery, they simply dismissed the opportunity to really understand the ramifications of the overlooked physics. They lacked curiosity. Moreover, they had their own priorities that had nothing to do with the completeness of the fundamental physics that they taught to all technical students within the university, physicists or otherwise. This is a theme I will return to. If the physicists don't believe there is unknown basic physics, then how could non-physicists react any other way than to deny the existence of such a circumstance?

Consequently, this book and the others in the set are a history of science research effort, a philosophy of science research effort, a physics research effort, and a technology research effort. These elements are woven throughout the four-volume set of books. The issue to me is that the quality and payoff for the research efforts within any specific areas depends on the knowledge and competency that a researcher brings to the research. Without a sufficient understanding of the science or technology being investigated, researchers are reduced to being true believers. If that is the case, we simply recording for posterity some sequence of events or observations made by others without understanding what we are recording for posterity. Often little original thought enters the quasi-journalistic approaches into researching and then scribing what has been observed, recorded, or discovered. My goal, on the other hand, was to learn enough about the subject matter to contribute to the body of knowledge and to make a qualified judgement relating to the subject as being good, bad, or ugly and then inserting myself into the discussion to "fix things." Plus, I have always embraced the challenges associated with problem solving.

We can summarize what was found and described in the earlier chapters by starting with listing the heuristics that were identified. These are listed in Table 5.1. The book subtitle only mentions one heuristic, which was Newton's accidental heuristic used in the gravitational law, where he assumed that the point-mass was representative of all mass distributions. But, as we have seen, there were other hidden heuristics that we use. Only the gravitational model was clearly associated with Newton, since only Huygens's pendulum anomaly suggested that dynamics using a center of mass might be a heuristic. However, to show that Newton's gravitational law was a heuristic, I needed to show that the point-mass concept was, in

Table 5.1

Heuristic
Point Mass/Charge
Center of Mass/Charge
Orbital Angular Momentum
Inverse-Square Force/Potential Laws
Multipole Expansion
Potential/Force Fields
Poisson's Equation

certain dynamic applications, also a heuristic, one that was unknown to Newton and to all the rest of us.

All the heuristics in Table 5.1 relate to having used the point mass and center of mass when there was a more rigorous method available, though most of these approaches and applications post-date Newton. However, these heuristics, while useful and generally accurate enough for all practical purposes, still represent shortcuts and simplifications. It may be alright for technologists and engineers to use these heuristics, given the accuracies and ranges of validity for these heuristics, but it is not alright that physicists use these same heuristics without testing them for validity over the range of experiments and models that are being developed. Worse, the heuristics have skewed the interpretation of the results of many observations that have been made.

For instance, we have clear analytical evidence that there are no such things as generalized test charges or test masses and fields as we currently understand them. There are no standalone fields, since we do not know a field is present or exists until we observe the behavior of a test mass or charge during an experiment. Not with standing that we know very little about the mass and charge distributions in subatomic particles, the results in this book clearly show that any non-point or non-corpuscular distributions yield results that are inconsistent with the point-mass, point-charge, and corpuscular assumptions.

The current use of point-mass or point-charge heuristics in the models for those listed in the table result in the derivative models, such as orbital models, also being heuristics and inaccurate over certain domains, typically for close-range interactions. Any model that uses a heuristic is itself a heuristic. We have also shown that the inverse-square or point-mass force law is a form that also holds for mutual electrical interaction, which makes our static electromagnetics heuristic physics. In so far as the analogies hold between Newtonian gravity and electrostatics, we have a range of new hidden heuristics that will also need to be explored to ascertain the ranges and limits over which electrostatics can be trusted.

Engineers, technologists, and scientists in all disciplines need to understand what may be heuristics that they have been using and that there may be occasions when the exact rigorous models must be used. But more egregiously, scientists in other areas of the physical sciences, such as in astronomy, cosmology, planetary and atmospheric physics, geophysics, and various other current and paleo-applications of these disciplines, need to recognize that they are using heuristics and that some of the unexplained observations in their respective disciplines may be explained not by positing new physics but by using rigorous models and physics rather than heuristic models and laws.

In the few examples given in this book, we have seen that our beliefs in certain physics may be entirely wrong and that we are, in addition, missing some possible technologies. The models for the fields within a sphere, for example, show that the forces are such that we may have phenomena we have not predicted, such as non-thermal pressure gradients driving flows of materials within volumes. If true, this has a profound impact on geophysics and stellar physics.

By and large, spheres show the least variance from the point-mass model in the external mutual forces with other spheres because of the three-dimensional symmetry and our use of uniform spheres. Naturally occurring spherical objects always have some deviations from being perfect and uniform spheres. Consequently, uniform spheres may be producing idealized and unrealizable mutual forces in the same way that the point-mass model shows idealized mutual forces. The oblate spheroidal shapes of rotating objects may introduce variances in system dynamics when the general Newtonian gravity model is used, which produces, over time, the types of systems we encounter in planetary, galactic, and cosmic distributions of mass.

The most important contribution of this book to physics is to show that we have erred in believing that a consensus among academic and practicing physicists represents the truth concerning the subject of the consensus. While we challenge and sometimes bitterly condemn many modern research efforts, as time passes we allow bad science to thrive. This is the converse of Planck's quote regarding the acceptance of new ideas. As time passes, we should become more skeptical of prior efforts and re-vet these efforts to ensure that prior generations of scientist did not accept and embrace false or incomplete science. The danger is that, otherwise, heuristics become rigid beliefs that skew the scientific and technological trajectories of various pursuits.

In the next book on relativity, we examine other assumptions, assertions, and beliefs and identify the consequences to technology of eliminating false and erroneous beliefs. And, taken a step further, we identify a potentially existential technology in the third book based on revisions and extensions in our interpretation of Newton's third law on action and reaction. It still exists, but it is not necessarily a totally reciprocal effect, and for an action there may not be exactly an equal and opposite reaction, though this is a systems consequence and not the exact consequence of Newton's laws of motion. Newton's dynamic laws are point-mass, simple point-force, and inertial laws, which we investigated for the third book for being heuristics or at least incomplete with regard to the impact of the linear laws within non-linear and non-inertial systems. The consequence was that we identified an unexpected asymmetry that we can exploit in identifying a new technology.

Appendices

Appendix 1—Orbital Angular Momentum

This is the topic that started the odyssey that led to the discovery that Newtonian physics was incomplete and incompletely understood. On a DARPA contract, I had been looking for signatures for satellites and re-entry vehicles. Signatures are specific measurements that allow an object to be identified or its dynamics to be recognized and modeled. Sensor systems used for acquisition and tracking of objects are often required to supply more thorough determination of what an object is and what its trajectory may be, depending on what the object is and what the mission scenario may be.

In performing such an analysis, I came across some anomalous results that appeared to be inconsequential to the goals of the contract. These results were relegated to an appendix in the final report, along with suggestions that they should have more thorough study. Those studies apparently never occurred until I began to question certain aspects of the physics that I had been taught and had successfully used.

However, upon re-examination of the anomaly, once I understood what I was modeling, I carried out some analyses that that led me to similar anomalous results in Newtonian gravity, which are discussed in Chapters 3 and 4 and in Appendix 2. None the less, because there is little practical fall out from "getting the orbital physics right," I relegated the details of the discussion to an appendix of this book. The phenomenon I will describe is, therefore, primarily of academic interest and represents a consequence of not doing the physics that is explicitly required and, instead, using a shortcut, which is Newton's point-mass approach to modeling the dynamics of an object.

Some angular momentum perturbations might be measured and, therefore, recognized as perturbations, and we will quantify some examples. However, some perturbations appear to only be important in collision dynamics, such as occurs in astrophysics or in nuclear and particle physics. None the less, the dynamics models for an object under the influence of a central force currently are missing some physics, specifically a complete description of the angular momentum. Now we can develop a model and quantify the degree of dynamic perturbation and its observability in more practical scenarios. The orbiting object

Appendix 1—Orbital Angular Momentum

will initially be identified as a sphere, though later we also look at cylindrical objects. In chapter 1 we discussed an orbiting dumbbell, and there is no reason to continue to refine that discussion.

What follows is presented in Chapter 1 and in first-year college physics books but it is repeated here because it is more convenient to establish a style of presenting this information for this book. The angular momentum of an orbiting object is given as $\mathbf{L} = \mathbf{r} \times \mathbf{p}$, where the bold letters represent vectors, $\mathbf{L}$ is the angular momentum, $\mathbf{r}$ is the vector distance of some mass moving with a momentum given by $\mathbf{p}$, and $\mathbf{X}$ is the notation for a cross product, which is the sine of the angle θ between $\mathbf{r}$ and $\mathbf{p}$.

When a spherical object with mass m is orbiting about or passing by some point under the influence of some force and at a distance Δ between the orbital point and the center of mass of the object, the typical approach is to use the center of mass as the location of the entire object's mass, m. Hence, the magnitude of the angular momentum is given as $L = \Delta\, m\, v\, \sin\theta$, where v is the linear velocity of the object's center of mass relative to some coordinate system in which Δ has specific coordinates. In polar coordinates, $v = \omega\, \Delta$, where ω is the angular rate of the mass with respect to the point defining the radius of curvature of the motion of the object. Also, in the center of mass, $\theta = 90$ deg and $\sin\theta = 1$. Therefore, $L_{cm} = m\, \Delta^2\, \omega$.

The angular rate is constant for the entire orbiting mass, whereas the velocity at any i^{th} mass element within the object depends on the magnitude of Δ_i. However, the direction of v is constant for all points within the object but the magnitude is not. When we look at any given mass element within the object and draw a vector from the orbiting center of curvature, which is defined below as the system barycenter, we have that the incremental i^{th} angular momentum of the ith mass element is given as $L_i = m_i\, R_i\, v_i\, \sin\theta_i$, the angular rate ω for the entire sphere is constant but v_i depends on the magnitude of R_i. For a sphere, we can use cylindrical or spherical coordinates defined around a coordinate axis given by the line joining the orbit's barycenter and the center of mass of the orbiting object.

We define all parameters, then, with respect to a sphere within a cylindrical coordinate system, as shown in Figs. A1.1 and A1.2. The total orbital angular momentum of the system is given by integrating all the incremental angular momentum components across the volume of the object, which in this case is a sphere. The most straight forward way to define the integral is to set up a coordinate system in which the z-axis lies along the vector Δ between the center of mass and the barycenter or orbital center of motion. This would be true for spherical and cylindrical coordinates. We then can find the cross product of a mass-element location with the velocity vector, which is now aligned in the y-axis direction, by using a 3x3 matrix to carry out the vector cross product component by component. The cross product gives the perpendicular component of one vector relative to another.

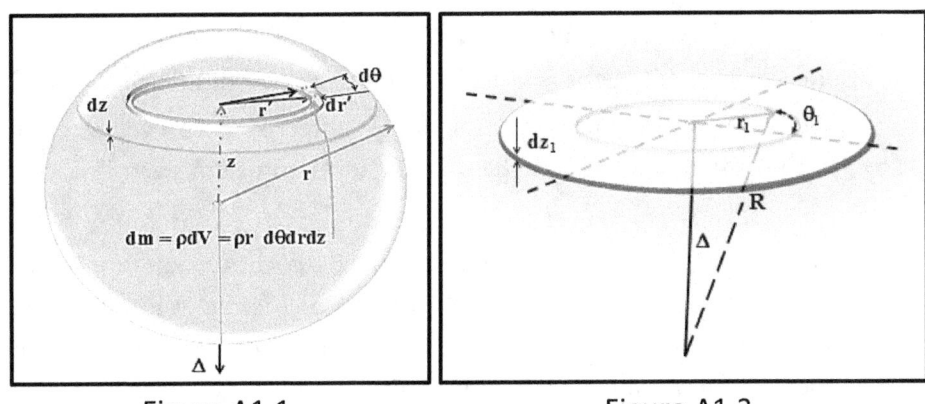

Figure A1.1   Figure A1.2

We can find either the component of the velocity vector perpendicular to a lever arm vector or we can find the perpendicular component of the lever arm vector relative to the velocity vector. The result is the same. We have for the angular momentum that $\mathbf{L}_i = \mathbf{R}_i \times \mathbf{p}_i$, where $\mathbf{R}_i$ is the vector from the barycenter of motion to the ith location within the sphere of radius r, $\mathbf{p}_i$ is the momentum of the ith incremental mass at location $\mathbf{R}_i$, and the cross product operator, $\mathbf{X}$, finds the sine of the angle between the two vectors. The cross product gives the projection of the length $\mathbf{R}_i$ perpendicular to the velocity $\mathbf{v}$. But the velocity $\mathbf{v}$ is given by the angular rate times the same projected perpendicular component of $\mathbf{R}_i$ relative to $\mathbf{v}$.

We find the perpendicular component of $\mathbf{R}_i$ by selecting the direction of $\mathbf{v}$ in our coordinate system and taking the cross product of $\mathbf{R}_i \times \mathbf{v}$. We use a coordinate system in which the vector Δ is in the $\hat{\mathbf{k}}$ or z-direction and only components of $\mathbf{R}_i$ perpendicular to $\hat{\mathbf{k}}$ contribute to the angular momentum. Forming the cross product as a matrix, we find $\mathbf{R} \times (\omega R_\perp \hat{\mathbf{k}})$, which is, in matrix form, given as:

Appendix 1—Orbital Angular Momentum

$$\begin{vmatrix} \hat{i} & \hat{j} & \hat{k} \\ -r'\sin\theta & r'\cos\theta & \Delta+z' \\ 0 & \omega R_\perp & 0 \end{vmatrix} = \omega R_\perp \{\hat{i}(-(\Delta+z'))+\hat{k}(r'\sin\theta)\},$$

where $\hat{i}, \hat{j}, \hat{k}$ are unit vectors in the x, y, and z directions, respectively. The components of $\mathbf{R}_\perp$ are given by $\hat{i}(-(\Delta+z'))+\hat{k}(r'\sin\theta)$ and the magnitude of $\mathbf{R}_\perp$ is given by $\sqrt{(-(\Delta+z'))^2+(r'\sin\theta)^2}$. In effect, we find $\mathbf{R}_\perp$ twice: once to define the velocity at each location, since the magnitude of v is $\omega R_\perp$, and a second time to find the perpendicular distance of $\mathbf{R}$ to each v_i at each location, so that the magnitude of the angular momentum at each ith location is $L_i = R_{i\perp}\, v_i = R_{i\perp}^2\, \omega$.

We can compare the above exact approach to the moment of inertia approach, in which $L = I \omega$ and I is the moment of inertia. When we have a spin axis within an object, the distance from that axis to any mass element is always perpendicular to the ith velocity at the ith mass element, so that the velocity of that element is always perpendicular to the line from the spin axis to the mass element.

If we were to just have a point-mass description, then the location of the center of mass in the same coordinate system would simply be $-\hat{i}\Delta$, using the same matrix approach as above. Therefore, the magnitude of the total point-mass angular momentum would be given as $L = \omega m \Delta^2$, where m is the total mass and the vector direction is in the $-\hat{i}$ direction, given the coordinate system we set up and the chosen direction for the orbital velocity. The next step is to determine if the discrete model requiring integration over the volume of the sphere is the same as the point-mass model.

The total magnitude of the angular momentum is found from evaluating the following integral:

$$L = \iiint \omega\rho(\Delta^2 + z'^2 + 2\Delta z' + r'^2 \sin^2\theta)\, r'dr'\, d\theta\, dz',$$

where the differential mass at any location is given in cylindrical coordinates as $\varrho\, r'\, dr'\, dz'\, d\theta$ and where ϱ is the local density, which need not be uniform across the sphere. In the integral, the variables r', z', and θ are local to the sphere, whose center of mass is Δ away from the center of motion. In addition, $r'^2 = r_s^2 - z'^2$, where r_s is the radius of the sphere and we perform the integration over r' and θ before performing the integration over z'.

Upon evaluation, we find that the function is analytical for the sphere. Using a unit sphere and normalizing the calculated angular momentum with the point-mass value at the same value for Δ yields the following curves, where Δ = d in

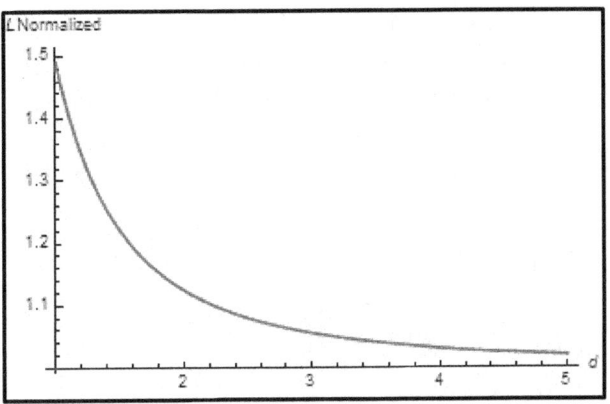

Figure A1.3—Orbital Angular Momentum (L) for a Unit Sphere normalized to the point-mass orbital angular momentum. Orbital range for Δ = d is given as multiples of the sphere radius and is 1 ≤ d ≤ 5.

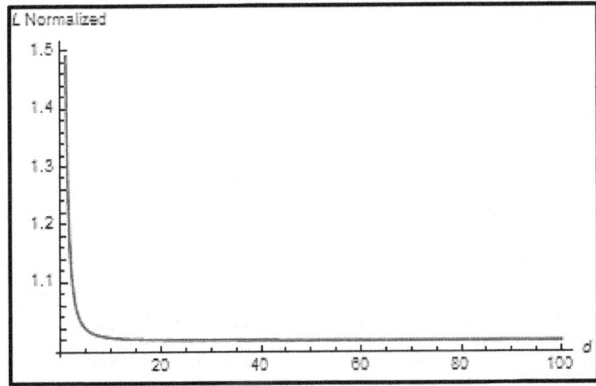

Figure A1.4—Orbital Angular Momentum (L) for a Unit Sphere normalized to the point-mass orbital angular momentum. Orbital range for Δ = d is given as multiples of the sphere radius and is 5 ≤ d ≤ 100.

the following two figures. Figure A1.3 is plotted for the range of values of *d* ranging from 0 ≤d ≤ 5 times the radius of the unit sphere, and Fig. A1.4 is plotted for the range of values of d from 5 ≤d ≤100 times the radius of the unit sphere. The center of mass angular momentum is unity for all values of d. The plots begin with the relative angular momentum calculated at d = 1, which is a point on the surface of the sphere.

The value of the angular momentum for a uniform sphere is L = 1.4 L_{cm} for d = 1, which is when the center of motion distance d equals the radius of the sphere…the sphere is pivoting about a point on its surface. In contrast to this value for the angular momentum, at d = 100 we have L = 1.00004 L_{cm}. In other words, the exact orbital angular momentum is always larger than for the point-mass model and approaches the point-mass model asymptotically from above as the value of d increases.

Figures A1.5 and A1.6 show the range of angular momenta for the moon and for Mercury in their orbits compared to L_{cm} (dotted line in figures). The origin of the extra angular momentum is a hidden one-rotation per orbit that all objects undergo. It can be visualized by observing the moon, which is locked to the Earth with one side of the moon always facing the Earth. In one orbit, the moon also makes one revolution on its axis. All objects experience this and it is well known. However, what is unrecognized is that the hidden singe extra revolution per orbit is actually a component of

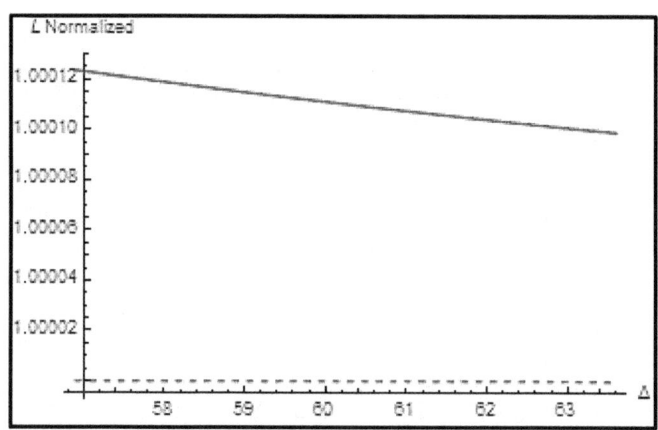

Figure A1.5—Orbital Angular Momentum for the moon compared to the point-mass orbital angular momentum (dotted line) over the orbital range 56.9 ≤ Δ ≤ 63.6 times the radius of the orbiting sphere.

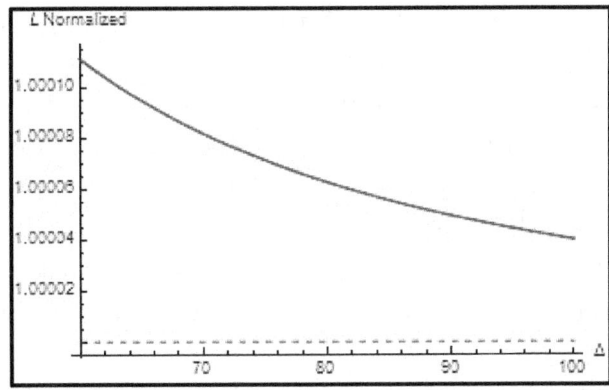

Figure A1.6—Orbital Angular Momentum for Mercury compared to the point-mass orbital angular momentum (dotted line) over the orbital range 66.1 ≤ Δ ≤ 100.3 times the radius of the orbiting sphere.

the orbital angular momentum. For both the moon and Mercury, the perturbation is on the order of ~ 10^{-4} of the point-mass angular momentum and is, as discussed below, a perturbation on the eccentricity of the orbits of these two objects.

We can also look at common artificial satellite shapes, such as spheres and cylinders, with the cylindrical satellite oriented with its axis pointing toward the Earth as well as pointing both along and across the direction of motion. We can then calculate the orbital angular momentum. In fact, if we have an analytical three-dimensional shape, we can calculate the orbital angular momenta for relative orientations of the objects, though enough symmetry must exist for the integrals to be carried out properly. As will be discussed in Appendix 2, even though we can write an integral for some models, the numerical integration of these transcendental functions can cause considerable difficulty.

For a uniform-density orbiting cylinder of radius r_c and length Λ, we first calculate the orbital angular momentum with the axis of the cylinder pointing toward the center of motion. Symbolically, the integration yields $L = L_{cm}[1 + (\Lambda^2 + 3 r_c^2)/12 \Delta^2]$, where $L_{cm} = \omega M \Delta^2$, which is the-point-mass orbital angular momentum and where Δ is the distance from the center of motion to the center of mass of the cylinder. The quantity $L_{cm}(\Lambda^2 + 3 r_c^2)/12 \Delta^2$ is the angular momentum of a cylinder rotating about a perpendicular bisector axis with angular rate ω. We see that this

Appendix 1—Orbital Angular Momentum

is simply the same single hidden rotation per orbit as identified for a sphere and this single rotation is coupled to the orbital angular momentum. We can continue and find the orbital angular momenta for the cylinder oriented with its axis along the direction of motion, which is $L = L_{cm} [1 + r_c^2/2 \Delta^2]$. For the direction of motion along a perpendicular bisector of the cylinder, $L = L_{cm} [1+(\Lambda^2 + 3 r_c^2)/12 \Delta^2]$. The results are consistent with the orbital angular momentum consisting of a point-mass orbital angular momentum plus an additional term for the single orbital rotation.

The magnitudes of the correction terms can be seen by inputting the cylindrical dimensions for the Hubble Space Telescope, which has dimensions $r_c = 2.1m$, $\Lambda = 13.2m$, and $\Delta \sim 6930km$ using the mean radius of the Earth. With these values, the perturbation on the orbital angular momenta is $\sim 2.25 \times 10^{-12}$ to $\sim 5 \times 10^{-15}$, depending on the dynamic orientation of the HST as it orbits the Earth. For a spherical satellite with a diameter of 4.2m, the perturbation is $\sim 4 \times 10^{-14}$. The perturbation increases the total orbital angular momentum but, since it only occurs with a period of the orbit, it is doubtful such a perturbation could be detected.

The orbital angular momentum is a key value in the orbital equations of an object. The angular momentum defines the orbital eccentricity. The eccentricity ε is given as $\varepsilon = \sqrt{1+kL^2}$, where k is a number associated with the orbiting object and system. If L increases, the eccentricity increases. For a closed orbit, $0 \leq \varepsilon < 1$, and when ε = 0 we have a circular orbit, and when ε > 0 we have elliptical orbits that become more elongated the larger ε becomes. When ε ≥ 1, the orbits are open and are trajectories and not orbits. Therefore, if the perturbations increase ε, the orbit becomes more elongated. If an object has additional rotation such that the instantaneous orientation of the object is changing, the orbital angular momenta are also fluctuating and, therefore, the eccentricity is also fluctuating, which means that there would be fluctuating changes in the trajectory of the object which would be like small ripples or scallops in the path.

Changes in the eccentricity also change the angular and radial speeds of the object. Thus, any fluctuations in the angular momentum would introduce orbital perturbations that would consist of periodic fluctuations in both the path and the velocity components of the object, which are the radial speed and position and the angular rate and position plus changes in the speed at periapsis and apsis. In other words, the orbiting object...or any object on any trajectory... would be subjected to micro-accelerations that would appear as long-term and repetitive variations in an accelerometer. Given the small size of the perturbation for man-made objects, it is doubtful that these would introduce any useful micro-signatures into the gross signatures of an object based on Doppler radar, lidar, or optical tracking systems.

However, we can carry out a similar analysis on planetary objects to confirm that these micro-accelerations are still too small to be a factor in the orbit, at least for their stable orbits. We need to prove that a large asteroid with an arbitrary shape might experience detectable orbital changes associated with our new understanding of the role of the non-point-mass orbital angular momenta on the orbital parameters for the object. A small object would experience no more drastic effects than artificial satellites. However, we need to look at a large object, such as a 1km diameter asteroid in the shape of a "potato" that is rotating at some rate or revolution per second end over end. If we were to see such an object at a distance the same as the Earth's orbit from the sun, we can hypothesize that the angular momentum variations might fluctuate over a range like that for a cylinder, ranging from the largest value to negligible during one rotation.

To make some estimates on the magnitude of the effects on an asteroid, we will simply use a scaled version of the HST using a length of 1km and a diameter of 0.5km. The distance from the sun is then 1.5×10^8 km for an Earth-orbit crossing object, though the object could be located any distance from Earth and would not necessarily be under the direct influence of Earth's gravity except as a small perturbation until the gravity influence of the Earth approaches that of the sun. However, depending on the distance from Earth, there could be some measurable perturbations from Earth's gravity, which is ignored in this analysis. Using the formula for the cylindrical perturbations, we find that the perturbation to the orbital angular momentum is $\sim 10^{-18}$, which is far smaller than for artificial satellites.

As the object becomes more directly subject to Earth's gravity, however, let's consider a scenario in which the distance of the asteroid from the Earth is three times the Earth's radius, a very close encounter. In this case, the perturbation could grow to become on the order of 10^{-9}, becoming even larger as the crossing distance becomes smaller. The rotation rate would be directly measurable and, consequently, the fluctuations should also be measurable using filters on the receiver's input processor. If the object is on a trajectory to just graze Earth, the angular momentum perturbation can be no greater than defined by the Earth's radius, which for even a 1km object would only be about four times the value calculated above. How knowledge of these small perturbations might affect our ability to track such objects would need to be the subject of further parametric analyses, though the consequences are likely too small to be consequential, unless the combinations of the perturbations in the angular momentum from solar and Earth influences lead to some resonance effects that re-defines the orbital stability. Additionally, the orbital stability may be affected by these small angular momentum perturbations, though stability theory may need to be modified to include the unknown but small perturbations.

Consequently, if there is any practical payoff for knowing that the angular momentum perturbation exists, we must identify some orbital parameter, including stability, that we can measure. Because the fluctuations are periodic and continual with a rate given by the object's rotation rate, while the rippling or scalloping effect on the object's dynamic path may be interesting, they would average out to zero in a complete orbit and should not affect the larger orbital dynamics. Those dynamics are defined by the long-term angular momentum perturbation as identified by the center of mass plus the shape-driven orbital angular momentum perturbation, though the fluctuating angular momentum my become important in determining orbital stability.

To find some measurable and meaningful perturbations, we need to understand what we mean when we say that the angular momentum of an orbiting object is conserved. For an elliptical orbit, the value of Δ, the distance of the object from the center of motion in the system, varies. Therefore, the angular momentum does appear to vary with time and it appears to fluctuate. None the less, the mean value of the orbital angular momentum is not varying and is constant. We see this by looking at the equation of motion for the orbiting object.

If we set up the Lagrangian, which is $L = T - U$, for an orbiting object, we can see where the missing angular momentum comes from. U is unchanged by the change in angular momentum but the kinetic energy now is $0.5\mu\dot{r}^2 + 0.5I\dot{\theta}^2$, where μ is the reduced mass, $\dot{r} = \dot{\Delta}$ and $\dot{\theta} = \omega$, where the symbols have been changed to coincide with the symbols used in Chapter 2, and I is the moment of inertia of the orbiting object which, for a rotating or tumbling object will fluctuate as the orientation of the object relative to the vector **Δ** varies. The variable change was made to coincide with more typical symbols used with the Euler-Lagrange equation.

As we saw for a cylinder, as the cylinder rotates at some rate ω_c, which is independent of the variable ω except for the single hidden rotation of the object in each orbit, the moment of inertia varies over a range of values that are dependent of the dimensions, mass distribution, and orientation of the object and the range of the object from the barycenter of the orbit, which is Δ. Now, if the object is rotating, what is the proper moment of inertia to use to find the kinetic energy, T, for the single hidden rotation? For the point-mass model, all we had to include was the single non-varying kinetic energy associated with the hidden rotation, but now we have an additional coupling of the native rotation rate of the object with the orbital angular momentum through the broadened definition of I for a rotating non-point-mass object.

For the case of a uniform sphere, the value of I was found to contain two terms: $I_{sphere} + \mu r^2$, where $I_{sphere} = 0.4 \mu r_s^2$, where we have used the reduced mass and where r_s is the radius of the sphere. Putting these into the kinetic energy term, T, gives us $T = 0.5\mu\dot{r}^2 + 0.5\dot{\theta}^2(\mu r^2 + 0.4\mu r_s^2)$, where the dot over a symbol represents differentiation with respect to time. We find the orbital equation by putting the Lagrangian into the Euler-Lagrange equation, which yields Lagrange's equations of motion for the orbit: $\{d(\partial L/\partial \dot{r})/dt\} - \partial L/\partial r = 0$ and $\{d(\partial L/\partial \dot{\theta})/dt\} - \partial L/\partial \theta = 0$. We won't solve this system of equations, but the orbital angular momentum falls out of this system of equations as $p_\theta = \partial L/\partial \dot{\theta}$. And, since U is a function of r and since $\partial L/\partial \theta = 0$, we have that $dp_\theta/dt = 0$ and p_θ is conserved. Reverting back to our earlier symbols of $r = \Delta$ and $\dot{\theta} = \omega$, we have that $p_\theta = \dot{\theta}(\mu r^2 + 0.4\mu r_s^2) = \omega(\mu\Delta^2 + 0.4\mu r_s^2)$.

In the point-mass equations, the term $0.4\ r_s^2/\Delta^2$ is missing. For the point-mass model, the time derivative of the angular momentum is $2\dot{\Delta}\Delta\omega + \Delta^2\dot{\omega} = 0$. By carrying out the algebra, we find that $2\dot{\Delta}/\Delta = -\dot{\omega}/\omega$. When the radial component of the orbital velocity is increasing, the angular rate is decreasing. This can be compared to the result when the perturbation term is included and we find that $2\dot{\Delta}\Delta\omega + \Delta^2\dot{\omega} + 0.4 r_s^2\dot{\omega} = 0$. Therefore, we have changed the relative components of velocity at each point in the orbit by a small amount, with the motion perturbations given by $\dot{\omega}/\omega = -2\dot{\Delta}\Delta/(\Delta^2 + 0.4 r_s^2)$. As Figs. A1.1 and A1.2 indicate, the perturbation only becomes significant when $\Delta < \sim 5$ radii of the smaller object, and the impact of the perturbation is more subtle for larger values of Δ. Even then, the objects must be of some similar sizes for the perturbation to be significant. We can also see that if there is some independent variation in the moment of inertia for an object, there will be fluctuations in the orbital angular momentum which, as discussed in Chapter 2, leads to fluctuations in the orbital eccentricity that appear to be scalloping motions of the object. For a near collision, the scalloping can become significant. If one were tracking such a near-collision scenario between comparably-sized objects, the dynamics when the impact parameter of the encounter is smallest can look decidedly non-Newtonian. And, given that natural objects can have an arbitrary shapes and mass distributions and arbitrary rotations, the magnitudes of the objects' moments of inertia at any instant can be significantly different from that of the simple shapes we have modeled.

What the analysis shows is that we should distinguish between the instantaneous angular momentum and the mean orbital angular momentum. Our current understanding of orbital dynamics is based on mean values and we have ignored…or been ignorant of…any intermediary fluctuations. Over time, the fluctuations average out, but on the short term, our measurement process could be capturing the contributions from the perturbations, if we could even measure these fluctuations, which would cause us to use incorrect values for the mean, thereby invalidating our trajectory estimates. Also, since natural objects are not uniform spheres, there will always be some orbital angular momentum perturbation for all objects, and those perturbations will fluctuate with the rotational rate of the object. Note again that for normal planetary orbits, the perturbations are likely undetectable. For collisions in any astrophysical or nuclear encounter, however, and when the separation distances are less than ~ 5 to 10 times the radius of the larger object, the perturbations can become significant.

Appendix 2—Newton's General Gravitational Model

We have already discussed the consequences of using the generalized Newtonian gravitational model for a variety of scenarios. In this appendix, we supply a few more details on the actual mathematical models used in analyzing the scenarios. It is clear that, the general model is likely the correct model to use and the point-mass model is a heuristic. What is less clear is, for many scenarios, what the general model is telling us, and it is not clear where this general model makes a difference in how we portray or model gravitational or electromagnetic interactions.

The broad question is, what new conclusions can we draw in a variety of fields in which the general model modifies how we think the universe works? On the one hand, the general model supplies a rationale as to why certain measurements are such as they are, such as why the measurement of the universal gravitational constant is the least accurate among the measurements for fundamental constants. On the other hand, the general model may supply a rational as to why, for instance, the motions of stellar objects within the galaxy are so unusual and difficult to interpret. But, on the third hand, from a practical perspective, the general model seems to supply few if any practical benefits over the use of the point-mass model.

To begin, we repeat the graphical representations of how the point-mass and general gravitational models are related. Figures A2.1a and A2.1b show how the general description (Fig. A2.1a) was modified by Newton to find the mutual force between a point mass and a spherical mass. When finding the magnitude of the force between two mass elements, we currently use $dF = \varrho_1 \varrho_2 G \, dV_1 dV_2 / |\mathbf{r}_{12}|^2$, where ϱ is the density and G is the universal gravitational constant. This is the conventional point-mass gravitational model. We use this model when we view the mutual force between extended object using the point-mass methodology. When we drop the explicit reference to a vector, we note that by symmetry the final gravitational force will be directed along the line connecting the centers of mass of the two objects, which for the point-mass model is the only vector that exists. For extended objects, we must perform a simultaneous integration across the two

Appendix 2—Newton's General Gravitational Model

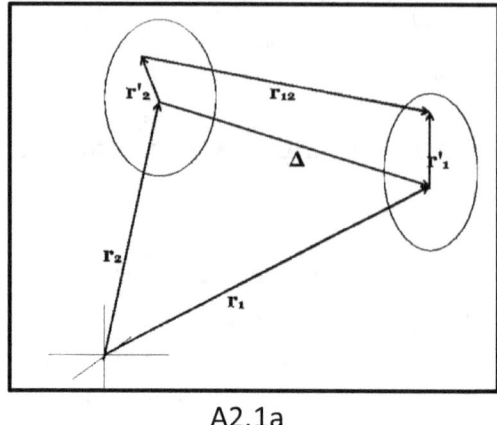

A2.1a

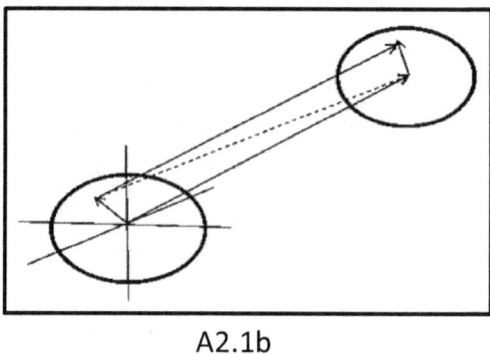

A2.1b

Figure A2.1—General vector description for Newtons' gravitational law (from Sokolnikov) which becomes the actual model used to find both the general solutions and Newton's point-mass solutions. (A2.1b)

volumes to find the mutual force between the two extended objects, where the magnitude of the vector $\mathbf{r}_{12}$ between mass elements is $|\mathbf{\Delta}+\mathbf{r}_2'-\mathbf{r}_1'|$.

Note that the standard point-mass form used in texts and solved by Legendre's method using modern calculus is $dF = \varrho \, M \, G \, dV_2 / |\mathbf{r}_2' - \mathbf{r}_1|^2$, where the mass of one sphere is reduced to a point mass of mass M located at $\mathbf{r}_1$ and we only integrate over the volume $\varrho \, dV_2$ for the other mass. However, the use of the vector with magnitude $|\mathbf{r}_{12}| = |\mathbf{\Delta}+\mathbf{r}_2'-\mathbf{r}_1'|$ is dictated by Corollary 2 in the *Principia*. It was never proven that the two forms are equivalent, that is, that $F = \int \varrho \, M \, G \, dV_2 / |\mathbf{r}_2'-\mathbf{r}_1|^2$ actually equals $F = \int \varrho_1 \varrho_2 \, G \, dV_1 \, dV_2 / |\mathbf{\Delta} + \mathbf{r}_2' - \mathbf{r}_1'|^2$. We can always use the simple point-mass model to model the mutual force between two point masses. These point masses are subsequently the incremental mass elements within each sphere.

However, we should also prove that the assertion that by symmetry the mutual force lies along the line connecting the centers of mass of the interacting objects. It may appear self-evident for a central force that this is a true statement, but the whole issue is that the use of the center of mass was a universal truth as we used it was also self-evident but has been shown not to be universally true. It is the non-linearity of the point-mass model that allows the general model to produce non-point-mass results.

Following Newton, we make the objects uniform spheres. Also, when we use the line Δ between the two centers of mass as a reference line, we have a natural cylindrical geometry with which to set up the integrals, as shown in Fig. A2.2. The new coordinate system is located at the center of mass of one of the spheres. Also, we are working only with magnitudes, though by symmetry the resulting force in along the vector Δ'. Using the cylindrical geometry described in Fig. A2.2, we find that the generalized model for Newton's gravity becomes:

$$F = \iiint \iiint \frac{G \rho_1 \rho_2 \Delta' \, r_1' \, r_2' \, dr_1' \, dr_2' \, d\theta_1 \, d\theta_2 \, dz_1' \, dz_2'}{\left(\Delta'^2 + r_1'^2 + r_2'^2 - 2 r_1' r_2' \cos(\theta_1 - \theta_2)\right)^{3/2}}, \qquad \text{Eq. A2.1a}$$

where $\Delta' = \Delta + z2 - z1$, which is the instantaneous separations of the two disk areas and is shown in Fig. A2.2.

Figure A2.2 depicts a cylindrical geometry, since for shapes such as cylinders this geometry is more flexible than the traditional spherical geometry used in the Legendre or multipole method. However, the spherical geometric description supplies other types of computational flexibility that cannot be duplicated with the cylindrical geometry. The spherical geometry allows us to find what happens as the density is give as a radially symmetric distribution within a sphere. We can also find the results for hollow shells interacting with other spheres, and we can also look at interior forces or potentials within spheres or shells, though we can do this with either coordinate system. Consequently, each geometry has its advantages. However, by using spheres in both coordinate systems interacting with each other, we also have a way of testing whether the deviations of the general model from the point-mass model are consistent or are possibly a result of computational issues. We will come back to this point later.

Appendix 2—Newton's General Gravitational Model

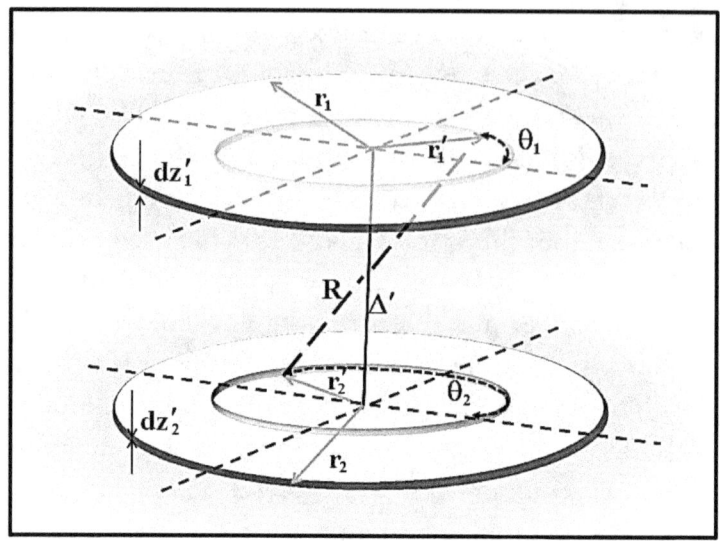

Figure A2.2—Cylindrical geometry used to model two extended objects

If we find the potential first and then find the mutual force, we would remove the term $\dfrac{\Delta'}{\left(\Delta'^2 + r_1'^2 + r_2'^2 - 2r_1' r_2' \cos(\theta_1 - \theta_2)\right)} = \cos\psi$ from Eq. A2.1a, where ψ is the angle between R and Δ' shown in Fig. A2.2. $\cos\psi$ is just the projection of the magnitude of the force between two elemental mass elements along the line joining the centers of mass of the two spheres, which is the direction of the mutual force. The integral for the mutual potential is:

$$U = \iiint \iiint \frac{G\rho_1 \rho_2 r_1' r_2' \, dr_1' \, dr_2' \, d\theta_1 \, d\theta_2 \, dz_1' \, dz_2'}{\left(\Delta'^2 + r_1'^2 + r_2'^2 - 2r_1' r_2' \cos(\theta_1 - \theta_2)\right)^{1/2}}. \qquad \text{Eq. A2.1b}$$

We can find the mutual force from the relationship $F = -dU/d\Delta = -\nabla U$, which is the negative gradient of the potential in the direction of mutual attraction and which is the direction of the line joining the centers of mass. This differential requires that the mutual force vector points along the line jointing the centers of mass of the two objects. If we find the potential for non-analytical functions, we would produce a graphical representation and then find the negative gradient of this curve to find the mutual force graphically. Consequently, for non-analytical functions, if we wish to find the mutual force, it is more convenient to produce the force results directly. For instance, Equations A2.1a and A2.1b have the appearances of elliptic integrals but are more complex. However, either integral is solved numerically using a program such as Mathematica® and easily graphed using the plotting functions available within Mathematica®. It is clearly better to find the mutual force numerically than to attempt to find this same mutual force using numerical integration, which is the only way to solve these equations.

The attractions between a variety of uniform spheres and differently shaped objects were calculated using a variety of spheres and cylinders, with the spheres both on the axis and on a perpendicular bisector of the cylinders. Cylinders were also evaluated against cylinders. In all instances, the results were not consistent with a point-mass model, and for cylinders the variations were sometimes extreme for near proximity of both cylinders. This latter result likely has consequences in the understanding the attained accuracy of measurements for the universal gravitational constant, the least precisely measure of all the fundamental constants. There is also the real effect that the orientation of an object on the surface of the Earth could supply what is perceived as a small…if it is even measurable…weight perturbation with orientation.

What the generalized Newtonian model shows is that any resultant mutual force depends on the relative sizes, mass distributions, and orientations as well as the separation distances between objects. Since the generalized model is not simply additive as is the classical Newton model, there is no compelling reason to finding the gravitational field of an object without referencing a test mass of some size, shape, and orientation, in complete contrast to the classical Newtonian results. However, the generalized model in some instances only supplies perturbations to the results of using the classical model, and most of the perturbations simply do not impact practical dynamics.

There are also likely consequences to certain electromagnetic forces, especially in electrostatics. The geometry depicted in Fig. A2.2 is suggestive of a capacitor. Calculations show that the mutual attraction or repulsion and the intervening field strengths are not the same as found by assuming the point charge approach is correct. We modify the gravity equation by replacing G by K, the Coulomb constant, and the volume increments with areal charge densities, which reduce the integrals by one dimension. In this form, the integrals are somewhat less complex than for the volume integrals. Once we have the forces, we can find the electric field, but again, the size of the test charge can become an issue in that if it is not a true point charge, the field is perturbed away from the point-charge value at any given location between the capacitor plates.

Appendix 2—Newton's General Gravitational Model

Rather than a uniform field within the volume of the capacitor, there are magnitude and directions variations across the plates, though the mean values may be very close to the total mutual force as found using traditional methods. However, edge effects are known to change the field strength near the edges of a capacitor. More analysis may show that such calculations are already capturing some of the effects predicted by the generalized Newtonian model for mutual attractions or repulsions.

Never the less, when we use test masses or test charges to find fields, we must also be aware that the actual dimensions of the test charges and masses will impact the analytical results. Fields do not exist as isolated entities, and they only exist as they affect the forces on finite charge and mass distributions. Finite sized spherical test masses or charges do approach the point-mass value for the interactions, though there seem to be statistical variations that may or may not be measurable. In fact, some experiments with capacitor plates only separated by atomic dimensions, as in the Casimir effect, do show some variances in the attraction or repulsion of the plates that is not consistent with the point-mass models. These variances may be a result of the charges of comparable size being in near proximity, even for putatively uncharged plates, and the small force variances may be predicted by the generalized Newtonian model and not be some other unknown forces.

If we recast the models for F and U using a spherical geometry, we have the following models in spherical coordinates:

$$U = \iiint \iiint \frac{G\rho_1 \rho_2 r_1'^2 r_2'^2 \sin\theta_1' \sin\theta_2' dr_1' dr_2' d\theta_1' d\theta_2' d\phi_1' d\phi_2'}{R}, \qquad \text{Eq. A2.3a}$$

where

$$R^2 = \left(r_2' \sin\theta_2' \cos\phi_2' - r_1' \sin\theta_1' \cos\phi_1'\right)^2 + \left(r_2' \sin\theta_2' \sin\phi_2' - r_1' \sin\theta_1' \sin\phi_1'\right)^2$$
$$+ \left(r_2' \cos\theta_2' + \Delta - r_1' \cos\theta_1'\right)^2,$$

and where the densities ρ can be functions of r, where the z-axis passes through the centers of mass of the two spheres, where φ is the azimuthal angle with range $0 \leq \varphi \leq 2\pi$, and θ is the declination angle between the z-axis and vector r' to any mass element within the spheres and has range $0 \leq \theta \leq \pi$. This is the way physicists define spherical coordinates, whereas mathematicians switch the roles of θ and φ. The separation between the two sphere centers is now simply Δ and the limits of integration are much simpler than in the cylindrical geometry. Nothing else, however, is simple.

If we let one of the spheres, say sphere one, have zero radius and mass M_1, then the above expression for the potential becomes $U = \iiint \iiint \dfrac{GM_1 \, \rho_2 r_2'^2 \sin\theta_2' \, dr_2' \, d\theta_2' \, d\phi_2'}{\left(\Delta^2 + r_2'^2 - 2\Delta r_2' \cos\theta_2'\right)^{1/2}}$, which is both the model Newton solved and the expression evaluated using Legendre polynomials. However, this is a much simpler integral than the Legendre's approach might suggest. Using methods taught in first year calculus, the integral for U can be evaluate analytically without resorting to the method using Legendre polynomials and without resorting to looking up integrals in tables of integrals. By extracting Δ^2 from the radical, the integral is evaluated as $\varrho \, V_2 = M_2$ and the expression for U becomes $G \, M_1 M_2 / \Delta$, using a simple variable-change approach, which gives Newton's point-mass result. However, we can also expand the radical sans Δ as an infinite power series, and when the integration is performed term by term on this series, all individual integrals have value zero except for the first term. The expansion into an infinite series is called the multipole expansion for reasons that are unimportant at this juncture and is the approach using the Legendre polynomials.

We can see that the general gravity model expressed in spherical coordinates appears to be much more complex than the comparable expression in cylindrical coordinates. Additionally, we are also unable to simply expand the integrand or, more specifically, the radical in the integrand as a simple power series. Thus, the multipole expansion is unique to the simple point-mass model and is only valid in the general model when the separation distances between spheres allow the general model to approach the point-mass model in the calculated values for the potential or the force.

Depending on the shapes of the interacting objects, we can also mix the coordinate systems. By this we mean that we can model a cylinder in cylindrical coordinates that is interacting with a sphere described in spherical coordinates. This is possible if we define a common z-axis and reference all coordinates within each object to this axis. The individual integral over one volume is independent of the integral over the other object.

If we were to evaluate U numerically, then we would need to generate a

Appendix 2—Newton's General Gravitational Model

curve from which to find the negative differential to find the mutual force. If we use the same approach with spherical coordinates as with cylindrical coordinates, we would need to find the cosine of the angle between the vector **R** defining the coordinates between the end points of the line connecting two arbitrary points in the spheres. The magnitude of this vector is just the value of the radical in the denominator of Eq. A2.3a, which is $R = \sqrt{R^2}$.

When we find the mutual force directly, we must include the cosine of the angle ψ between R and the z-axis, which is given as $\cos\psi = (r_2' \cos\theta_2' + \Delta - r_1' \cos\theta_1')/R$ in spherical coordinates and where by symmetry the mutual force is in the z-direction, which is along the vector **Δ**. Thus, the general mutual force is given by $dF = G\, dm_1\, dm_2 \cos\psi/R^2$. Written out completely, we have:

$$F = \iiint \iiint G\rho_1 \rho_2 (r_2' \cos\theta_2' + \Delta - r_1' \cos\theta_1')\, r_1'^2 r_2'^2 \sin\theta_1' \sin\theta_2'\, dr_1'\, dr_2'\, d\theta_1'\, d\theta_2'\, d\phi_1'\, d\phi_2' / R^3,$$

Eq. A2.3b

where R in spherical coordinates was supplied previously.

It is clear from the observation of the mutual force expression that Newton would likely not have been able to evaluate Eq. A2.3b using the methods at his disposal. Consequently, reducing one distributed mass to a point mass made the intractable tractable. On the other hand, given Newton's brilliance, he might just have come up with a way to solve this relationship had he thought it was necessary to do so.

To compare how U and F vary relative to the Newtonian values, we normalize each of the above integrals by the value of the Newtonian values for U and F supplied at the same separations Δ. From the point-mass approach, there are no volume integrals and we only use the mass for each sphere. To keep the sizes of the spheres consistent we do calculate the volume of the spheres, though, and the densities cancel between the two expressions. On the other hand, for a radial density profile, we do need to keep the functional form for the radial profile in the new general expression, which further differentiates the results from using the general model versus using the point-mass model.

Graphs were supplied in Chapters 3 and 4 for various scenarios. I used hundreds of scenarios to get a feel for how the mutual force would vary as size ratios, ranges, sampling granularities, and precision requirements were varied. From these some general conclusions were possible. Still, the calculations are meant to solve specific problems or to identify behavior in specific scenarios.

Ultimately, the general conclusions were that the point-mass force simply fails to capture the degree of variability present in the mutual force. It is not clear, however, how some of these variations might impact dynamic calculations.

What we have, then, is that the point-mass approach to the mutual gravity model is inherently smooth with no granularity in the calculated data points. The general Newtonian model, on the other hand, under certain scenarios…but not all…produces a much more variable data set, a characteristic of mutual interactions that is totally hidden without using the general model. However, the variances within the calculated data are often small and may simply be associated with just that, variances in the accuracy within the calculation for any given point.

References

References

arXiv.org—Archive maintained by Cornell University Library of mostly physics and mathematics pre-prints of papers, many of which are never refereed or published and are of mixed difficulty and quality.

URL= http://arxiv.org

Brillouin, Leon, *Relativity Re-examined*, Academic Press 1970

Chandrasekhar, S., *Newton's Principia for the Common Reader*, Clarendon Press: Reprint edition (June 12, 2003).

Dover Press—Affordable books on science and mathematics at many levels of difficulty, many of which were classics in their time. Searches on topics such as applied mathematics, analytical mechanics, and relativity will supply a lifetimes' reading in these subjects.

URL= http://store.doverpublications.com/

Famaey, Benoit and Binney, James, "Modified Newtonian dynamics in the Milky Way," Oxford Journals/Science & Mathematics/MNRAS/Volume 363, Issue 2/Pp. 603-608(Monthly Notices of the Royal Astronomical Society)

URL= https://mnras.oxfordjournals.org/content/363/2/603.full

Feynman, R., Leighton, R., and Sands, M., *The Feynman Lectures on Physics*, Addison Wesley 1963. On-line version available from CalTech:

URL = http://www.feynmanlectures.caltech.edu/

References

Frank, Adam and Gleiser, Marcelo, "A Crisis at the Edge of Physics," New York Times, Grey Matter, June 15, 2015

URL= http://www.nytimes.com/2015/06/07/opinion/a-crisis-at-the-edge-of-physics.html?smtyp=cur

Glashow, Sheldon, "The Errors & Animadversions of Honest Isaac Newton"

URL= http://www.iec.cat/butlleti/pdf/90_butlleti_sheldon.pdf

Goldstein, H., Poole, C., and Safko, J., *Classical Mechanics*, 3rd Ed., Addison Wesley 2000

Google Scholar—Indexed Search Engine for most science, mathematics, and technology topics, uses a familiar interface, and is often integrated with a university's own library catalog for rapid search and acquisition of electronic versions of scholarly works.

URL= https://scholar.google.com/

Gordin, Michael, *The Pseudo-science Wars*, The University of Chicago Press, Chicago 2012

Kleppner, D. and Kelenkow, R., *Introduction to Mechanics*, McGraw Hill 1973

MOND (The MOND Page)

URL= http://www.astro.umd.edu/~ssm/mond/

Newton, Isaac

URL= http://plato.stanford.edu/entries/newton/

Popper, Karl

URL= http://plato.stanford.edu/entries/popper/

Schaum's Outlines

URL= https://en.wikipedia.org/wiki/Schaum%27s_Outlines

Smith, George, "Newton's *Philosophiae Naturalis Principia Mathematica,*" *The Stanford Encyclopedia of Philosophy* (Winter 2008 Edition), Edward N. Zalta ed.

URL=http://plato.stanford.edu/archives/win2008/entries/newton-principia/

Sokolnikoff, I.S., *Tensor Analysis*, 2nd ed. (John Wiley & Sons, Inc. ,1964) pp. 259-263 (There appears to be a reprint available from Krieger Publishing Co. (October 1990).)

Thornton, S. T. and Marion, J. B., *Classical Dynamics of Particles and Systems*, 5th ed., Cengage Learning India 2012, pp. 422-424.

Weinberg, Steven, *To Explain the World: The Discovery of Modern Science*, Harper Perennial; Reprint edition (February 9, 2016)

Weinberg, Steven, Excerpts from "The Revolution that Didn't happen," *New York Review of Books*, Vol XLV, Number 15 (1998)

URL=http://www.physics.utah.edu/~detar/phys4910/readings/fundamentals/weinberg.html#back3

www.ingramcontent.com/pod-product-compliance
Lightning Source LLC
Chambersburg PA
CBHW062325220526
45469CB00008B/2621